Archaeoastronomy and the Roots of Science

AAAS Selected Symposia Series

Published by Westview Press, Inc.
5500 Central Avenue, Boulder, Colorado

for the

American Association for the Advancement of Science
1776 Massachusetts Ave, N.W., Washington, D.C.

Archaeoastronomy and the Roots of Science

Edited by E. C. Krupp

AAAS Selected Symposium **71**

AAAS Selected Symposia Series

This book is based on a symposium that was held at the 1980 AAAS National Annual Meeting in San Francisco, California, January 3-8. The symposium was sponsored by AAAS Sections D (Astronomy) and H (Anthropology).

Published in 1984 in the United States of America by
Westview Press, Inc.
5500 Central Avenue
Boulder, Colorado 80301
Frederick A. Praeger, President and Publisher

Library of Congress Catalog Card Number: 83-50763
ISBN: 0-86531-406-3

Printed and bound in the United States of America

About the Book

Archaeoastronomy is a rapidly developing interdisciplinary inquiry into the minds of our prehistoric and ancient ancestors, one that attempts to reconstruct the ways in which early peoples made use of the sky and its significance to them. Astronomy appears to be a fundamental component of culture, making the scope of archaeoastronomy worldwide. This book, reviewing recent research, includes new material on the megalith builders in Western Europe, North American Indians, the literate civilizations of Mesoamerica, the Andean culture of the Incas, and the Egyptians, among others. Recent discoveries and controversies are highlighted, and the relationship of ancient sky-watching to the development of true science is explored. This is not a tracing of the historical thread leading from pre-science to science; rather it is a cross-cultural search for those elements of prescientific activity that might give us a hint of the process of evolution of our own relationship with the sky.

About the Series

The *AAAS Selected Symposia Series* was begun in 1977 to provide a means for more permanently recording and more widely disseminating some of the valuable material which is discussed at the AAAS Annual National Meetings. The volumes in this *Series* are based on symposia held at the Meetings which address topics of current and continuing significance, both within and among the sciences, and in the areas in which science and technology impact on public policy. The *Series* format is designed to provide for rapid dissemination of information, so the papers are not typeset but are reproduced directly from the camera-copy submitted by the authors. The papers are organized and edited by the symposium arrangers who then become the editors of the various volumes. Most papers published in this *Series* are original contributions which have not been previously published, although in some cases additional papers from other sources have been added by an editor to provide a more comprehensive view of a particular topic. Symposia may be reports of new research or reviews of established work, particularly work of an interdisciplinary nature, since the AAAS Annual Meetings typically embrace the full range of the sciences and their societal implications.

WILLIAM D. CAREY
Executive Officer
American Association for
the Advancement of Science

Contents

About the Editor and Authors

E. C. Krupp, *director of Griffith Observatory in Los Angeles, California, has specialized in archaeoastronomy, ethnoastronomy, and the astronomical components of belief systems. He has examined or visited 400 ancient and prehistoric sites around the world, and he is the author or editor of numerous publications, most recently* Echoes of the Ancient Skies *(Harper and Row, 1983). In 1978 he was awarded the American Institute of Physics/U. S. Steel Award for Best Science Writing for his book* In Search of Ancient Astronomies *(Doubleday, 1978).*

Anthony F. Aveni *is Charles A. Dana Professor of Astronomy and Anthropology at Colgate University in Hamilton, New York. He is the author or editor of more than fifty books and articles in his field, among them* Native American Astronomy, Archaeoastronomy in Pre-Columbian America, *and* Skywatchers of Ancient Mexico *(all published by University of Texas Press, 1975, 1977, and 1980, respectively). In 1982 he was chosen National Professor of the Year by the Council for the Advancement and Support of Education.*

John B. Carlson *is founder and director of the Center for Archaeoastronomy, affiliate assistant professor of astronomy, and adjunct assistant professor of anthropology at the University of Maryland, College Park. A specialist in extragalactic astronomy and archaeoastronomy, he has done field research at more than 100 sites in the southwestern United States and Central America, with special emphasis on the Maya civilization. Currently he serves as editor-in-chief of* Archaeoastronomy, *associate editor of* Current Anthropology, *and advisory editor of the* Journal of the History of Astronomy.

Ronald Hicks, *associate professor of anthropology at Ball State University in Muncie, Indiana, has done research on prehistoric archaeology and the folklore associated with prehistoric earthworks in the British Isles and the midwestern United States. He is a fellow of the American Anthropological Association, the Royal Society of Antiquaries of Ireland, and the Society of Antiquaries of Scotland.*

Travis Hudson, *curator of anthropology at the Santa Barbara Museum of Natural History in Santa Barbara, California, has done work on Native American astronomy and archaeoastronomy, especially in California. He is the author of more than eighty publications, including* Crystals in the Sky: An Intellectual Odyssey Involving Chumash Astronomy, Cosmology, and Rock Art *(with E. Underhay; Ballena, 1978), and he was a Science Foundation Fellow and a National Academy of Sciences exchange scientist to the U.S.S.R.*

Archibald S. Thom *is honorary retired senior research fellow in the Department of Aeronautics and Fluid Mechanics at the University of Glasgow, Scotland. An engineer by training, he has also done research and written extensively on prehistoric sites in the British Isles. His books include* Megalithic Remains in Britain and Brittany *(with A. Thom; Oxford University Press, 1979) and* Megalithic Rings: Plans and Data for 229 Monuments in Britain *(with A. Thom and A. Burl; BAR, 1980).*

Acknowledgments

I am most grateful to all of the participants in this project for the vital commodity all possessed in abundance: patience. When it became apparent that both usual and unusual demands on the resources of Griffith Observatory would slow development of the final camera-ready copy, Joellen M. Fritsche, Associate Editor in the American Association for the Advancement of Science Publications Office, adopted a tolerant and understanding approach toward the delays. Her continuing support of the project and gentle prodding kept my spirits up and brought this effort to its earliest possible completion.

Similarly, the equanimity of the other five contributors greatly enhanced the production of this book. Their courtesy and professionalism made my job easier, and I am in their debt.

Because this volume evolved out of an AAAS symposium, the book would not have materialized had the symposium itself not taken place. It was organized for the January annual meeting in San Francisco in 1980. The symposium was possible only through the financial support of several institutions which subsidized the travel expenses for several of the participants. Mr. Robert L. Merritt secured a contribution from the Lloyd Foundation, which has supported research in archaeoastronomy before. Dr. John Carlson also was able to obtain a grant for the program from the University of Maryland. Dr. Priscilla Reining, representing the AAAS Section on Anthropology, provided additional financial assistance, and finally, Dr. Donat Wentzel, representing the AAAS Section on Astronomy, committed a substantial sum so that the symposium could take place. Indeed, Dr. Wentzel is very much responsible for the existence of this book, for he originally proposed the idea of holding an AAAS Symposium on archaeoastronomy and invited me to organize it. When he realized that the participants would have no independent sources of travel money, he took the initiative to secure the necessary funds. Your reading these words now is testimony of his success.

Certainly Griffith Observatory, Department of Recreation and Parks, City of Los Angeles, must be regarded as a coproducer of the symposium as well as of this book. I undertook the project as part of my duties at the Observatory with the approval of Mr. John Horan, Assistant General Manager, in charge of the Department's Scientific and Educational Branch. In addition, Griffith Observatory accepted the responsibility to produce the entire camera-ready manuscript. This meant preparation of many of the line drawings, diagrams, and photographs. Mr. Paul Roques and Mr. George Ewell Peirson did all of the necessary photocopying and printing. Observatory artists Ms. Helen Jorjorian and Mr. George Ewell Peirson did some of the drawings. And Mr. Joseph Bieniasz, Civil Engineering Drafter on the Observatory technical staff, executed many fine pen-and-ink diagrams.

Typing the manuscript to the standards of camera-ready pages was no small task either. Ms. Andrea Barkan, on the Observatory clerical staff, saw this demanding chore through from beginning to end and remained cheerful, dedicated, and accurate despite the other distractions and responsibilities of her job.

To all of those mentioned, then, I express my heartfelt thanks. All of them watched the progress of this work, first as one lunation followed another, and then as the sun itself completed a few seasonal rounds, never losing faith that the finish of this project was part of the natural order of things.

E. C. Krupp

1. As the World Turns

If careful observation of the world around us counts as science, then there is no doubt that our ancient and prehistoric ancestors were scientists. They would not have survived without a detailed knowledge of the environment.

Of course, there is much more to science than looking at the world meticulously and systematically. How we handle the information we gather and what we make of it are key components of the scientific process. But science does begin with an alert and curious eye. In antiquity, that eye was often turned to the sky. There it plucked patterns, snared cycles, and appropriated order from the dependable, repetitious movement of celestial objects. They became references in the landscape of time and space. From them the ancients fashioned systems of timekeeping and direction.

Calendars, clocks, and direction are part of our common heritage from the past. We don't have to reinvent them. They are foundations for organized human interaction, so necessary and so regularly used that we take them for granted. But our ancestors didn't take them for granted. To them, the sky was alive and powerful. The sun, moon, and stars did not answer to earth. In heaven, they were free from manipulation and the will of others. The sky's theatrics oriented the minds of our ancestors; the sky put meaning into the world. It made the world comprehensible.

Astronomy, then, was incorporated into numerous aspects of life, and the interdisciplinary study of ancient astronomy allows us to see how the sky was interwoven into traditions, temples, tombs, and tales. Astronomy was a fundamental component of culture.

Traditional peoples throughout the world and in every era have watched the sky. We can hope to understand how their relationships with the sky affected their lives only by approaching their views of nature on their terms. We are dealing here with something much broader than the origins of science. The way our ancestors interacted with the sky--the methods they used to watch it and the symbolic cultural expressions of what they saw there--immerses us into their belief systems.

Like ourselves, our ancestors fabricated a vision of the cosmos and tried to understand their place in it. Their attempts to come to terms with the universe were guided and inevitably sustained by their perception of its structure. For them, this structure was the cosmic pattern of time and space revealed by the cyclical dramas played out overhead.

If the earth did not turn on its axis each day and orbit the sun once a year, we would have a very different feeling about the orientation of the landscape and the passage of time. And were there no moon to parcel out the days in one-month packages, tied by the ribbon of its phases, there would be virtually nothing in the sky to provide us with a sense of ordered time and primary direction. The sky would be a very different place, with no beginnings, no endings, and no renewals. The world would be still, unchanging.

Instead, the sky is in constant transformation. Day and night exchange places with endless oscillation. The seasons pass on a continuous loop from one to the next until the first comes to the front of the line again to start the circuit once more, while the sun and the stars keep in pace with it. The moon keeps up a rhythm in movement and shape that allows anyone who stops to watch it to see a never-ending string of cycles.

The astronomical phenomena that supplied our ancestors with cyclical time told them the same story again and again:

Things come.
Things go.
Things come again.

This pattern was paralleled in the seasonal cycles of vegetation. This is the great pattern of life, and the comings and goings of celestial objects were associated with it. In these terms, the story was more compelling:

Things are born.
Things grow and mature.
Things die.
Things start over again.

It is still a simple story, one that mirrors our own experience. But it doesn't sound like science.

The sun doesn't really die at every sunset. It doesn't expire in the dead of winter at every winter solstice. The moon doesn't really die when it seems to be pared--through one waning phase after another--until there is no moon at all. The stars don't really die when they set below the western horizon. But we have said they do. These are stories we have told about sky, metaphors that helped us put the things we saw happening into language that matched the things that mattered to us most. Many things do die in the winter, and the world comes to life in the spring. In more tropical environments the basic pattern is the same, but it is linked with the cycle of rain and drought. Our ancestors saw a connection between the earth's seasons and the sun's annual movement. Therefore, they associated the celestial events overhead with the terrestrial events at their feet.

Metaphor was the link between these two realms, and metaphor is just a way of describing--of emphasizing--a relationship between things. Celestial metaphors have been expressed in many ways by traditional peoples: in myth, in systems of kinship, in the authorization of a ruler's power, in monumental architecture, and in many other cultural institutions. These embody the imagery of the sky because the sky is a reservoir of power. Its order provides a model for the order we must impose on so many areas of human activity.

For example, the emperor of China was regarded as the counterpart of the north celestial pole

The place of the emperor in China was the Hall of Supreme Harmony, the heart of the Imperial Palace in Beijing. This building straddles the primary axis of the capital, a north-south line that structures the entire city. Altogether the palace was called "the Purple (Polar) Forbidden City." It complemented the celestial zone around the pole of the sky, which was known as "the Forbidden Purple Palace." The emperor was the pivot, the hub, of terrestrial affairs, just as the pole was the spot around which all the sky turned. And in this capacity the emperor was the conduit that carried sacred cosmic order to earth. (Robin Rector Krupp)

because he, like the pole, was the source of stability and order. He was the fixed spot around which society turned, the center of the world, and his mandate to rule derived from the one fixed spot in the sky, the center of its diurnal motion. As the world turns, the pole stays put and appears to be the hub of the fundamental cycle of the world: the transformation of day and night.

Equating the celestial pole with the emperor was more than evocative allusion. It was symbolic language that established a relationship between the world, the emperor, and the sky with economy and precision.

Our brains help us get along in the world by collecting, categorizing, and correlating information, but if we really had to contend with all of the data our senses collect, we would be lost in a blizzard of details. Instead, most of the time, we focus on those things that are important, useful, and reconcilable with previous experience. Our brains are filters. They edit the world for us and so make it possible for us to survive in it. Our brains process the information, and symbolic, metaphorical thought allows us to manipulate it and communicate it.

We look, then, for pattern and order, and they show up in the sky more explicitly and more forcefully than in any other realm in our environment. But the sky is the scene, now and then, of departures from order: the violence of the storm, the unexpected appearance of a comet, the erratic path of a fireball, the devastating and sudden aspect of an eclipse. But these intrusions of chaos in an otherwise orderly kingdom pass. Order is always restored, and it is the reliable, familiar, cyclical behavior of the sky that makes it--for traditional peoples--a model of the structure of space and time.

Science uses models, too, and for the same reason: to express relationships. The primary difference between the use of models or metaphors in the scientific process, and their use by traditional peoples is the means by which we abandon the model.

It is possible, of course, to be mistaken about some of the relationships we believe to be true. In our culture, we have found it useful--a selective advantage--to refine our understanding of nature by discarding metaphors, models, or theories that don't stand up to scientific scrutiny. This systematic testing of our metaphors against all pertinent data is, to a great extent, just what we mean by science, and it is different from what we usually encounter in the belief systems of traditional peoples.

This book explores the celestial ingredients in several ancient belief systems and probes the boundary between them and science. The book began as a special symposium with the same title and organized for the 146th National Meeting of the American Association for the Advancement of Science and presented in San Francisco 6 January 1980. Because the theme of the National Meeting was "Science: Our Common Heritage," a lecture series on archaeoastronomy and the roots of science was especially appropriate. The presentations of the symposium were, in part, transformed into the six articles included here, and each of the authors demonstrates that the peoples considered here all watched the sky and paid attention to it.

The evidence for ancient astronomy is not always explicit or complete. In some cases, we are dealing with peoples who had no written language and so left no written account of their discoveries and concerns. But in the first paper, on California Indian traditions, Dr. Travis Hudson, Curator of Anthropology at the Santa Barbara Museum of Natural History, lets us see the richness of the ethnographic record. This record is extremely important, for it confirms that calendars, sky lore, astronomically inspired ritual were by no means the exclusive province of settled, agricultural communities.

Sometimes, however, the evidence is less clear. Some alignments in the neolithic and bronze age stone monuments of prehistoric Britain have been recognized as astronomical for many decades. Yet, only the old stones remain to tell us the story. Interpretations of this archaeological evidence is controversial. Dr. A.S. Thom is one of the investigators who has worked extensively out in the field for many years. He is very close to the problem of "megalithic science," and his paper here summarizes the case for precise solar and lunar observatories among the megaliths.

Celestial metaphors rest upon concrete foundations of careful, systematic observation. Observatories like this, built by astronomer Guo Shou jing at Gao cheng, were sources of the calendrical and astronomical information that helped order Chinese life. The emperor was responsible for acquisition of this information, and it, in turn, helped provide his mandate to rule. (Robin Rector Krupp)

<u>Maat</u> was the right order of things, the congruence of all components of the cosmos. It was a gift from the gods in the sky because they, as cyclical celestial objects, provided pattern and order on the stage overhead, and their performances made life possible here on earth. Shown here as a goddess in the Tomb of Seti I, Maat has wings, and they confirm her link with the sky. (E.C. Krupp)

Prehistorian Dr. Ronald Hicks of Ball State University, takes a somewhat different point of view on stone and bronze age astronomy in his contribution to this collection. Although the conflicting opinions evoked by the megalithic alignments emphasize the limits of our knowledge of the past, we can in any case be certain that the events in the sky were an important part of prehistoric life.

Dr. John Carlson, Director of the Center for Archaeoastronomy, seems, by the title of his paper, to have a less difficult task. His examination of Mesoamerican astronomy focuses on the native texts. Unlike the previous researchers, Dr. Carlson has a written record to help reveal the goals, techniques, results, and meaning of skywatching in ancient Mexico. He shows us, however, that interpretation of this record is not as straightfoward as we might guess. A true appreciation of its significance requires an understanding of its symbolic content.

The Mesoamerican peoples observed the sky from a tropical latitude. Dr. Anthony F. Aveni's extensive fieldwork in Mesoamerica has documented some of the consequences of the tropical view of the cosmos in the monumental architecture of the Maya and the other peoples of ancient Mexico. Dr. Aveni is Professor of Astronomy and Anthropology at Colgate University, and he has been responsible for promoting genuine interdisciplinary study of an ancient astronomy. His paper here incorporates some of his work in Peru and embraces the broader issue of the effect of latitude on astronomy and the belief systems that attend it.

Finally, the last article takes a brief look at the astronomy of ancient Egypt. We can be certain in some sense the roots of science must have penetrated the black fertile soil of the Nile, for the calendar and timekeeping conventions devised by the Egyptians eventually evolved into our own. Here, too, we have a written record to guide us, but the record, again, is not explicit. The Egyptian material reflects the astronomical systems that were in place, but very little of it comes to us straight from the files of the ancient Egyptian astronomers.

Despite the obstacles in the evidence, systematic observation of the sky has, we discover, a respectable antiquity. It was a fundamental activity--something that just about everybody did. In a quest for the roots of science, any sign of systematic observation suggests we are on the right track. But if we believe that our evaluation of the precision of megalithic observatories, the reliability of the Egyptian calendars, or the accuracy of the computational techniques in the Maya codices will put us face-to-face with science as we practice it today, we are bound to be misled. The skywatchers of antiquity were specialists in the sacred, not in science. This does not mean they were careless. Most of them knew the sky intimately. Their observations of the heavens supplied them with a sense of cosmic order. This is not just an intellectual insight; it is an emotional response as well. And it is this sense of cosmic order that was at the center of their concept of the sacred.

Recognition and celebration of cosmic order animated the religions of our ancestors. The rightness of things--or Maat, as the Egyptians deified it--was the primary gift bestowed by celestial gods. Guaranteeing the world's stability, well-being, and congruence, was their responsibility. The ancients' astronomies told them about gods in the sky, not about objects in space.

It is, however, with such an eye on the cosmos that science begins--curious, disciplined, and imaginative. Our ancestors harvested powerful and effective symbols from the sky. These resonated with their brains and through pattern and order made the universe make sense. Through this process the ancients fashioned a cosmology, a world-view. And a world-view doesn't have to be scientific to be useful.

Today cosmology is scientific, but our need for it is hardly different from our ancestors. The ancient astronomies may not be the same thing as modern science, but they show us how important understanding where we fit into the cosmos is to us. We, like our ancestors, draw pictures of the universe and try to figure out where in the cosmos we are.

2. California's First Astronomers

In the global search for understanding the origins of complex astronomies--and therein the roots of science--among ancient peoples, most research has focused upon food producers, particularly those who achieved a degree of civilization and left impressive monuments or written records. With the exception of Marshack's (1972) work, which concerned the European Paleolithic, food gatherers have been excluded from this search for a variety of reasons. The most common is the general acceptance among scholars that these people would have had little need to be concerned with understanding or predicting changes taking place in the heavens. Indeed, over time this assumption has become so deeply rooted in our conception of human accomplishment that it is simply accepted as "fact." If people lacked crops or flocks, they also lacked a sophisticated awareness of the heavens. Expressed another way, astronomy and calendrics were the inventions of food producers.

When applied to California's food gatherers this assumption is untenable. The ethno- and archaeo-astronomical data presented in this paper reveal that California's native inhabitants had a profound and long-standing involvement with the heavens, one certainly far more sophisticated than previously realized. My purpose is to present a general picture of this complex body of knowledge among California's hunter-gatherers, and to offer a strong challenge to the assumption that hunter-gatherers did not achieve a relatively sophisticated astronomy.

There is another important purpose to this paper. Because the evidence supports the existence of a sophisticated knowledge of astronomy on the part of these food gatherers, it is reasonable to infer this was also true of other ancient hunter-gatherers, particularly those in regions of the world where a transition from food collection to food production took place (e.g., Mesopotamia, Egypt, China, Mesoamerica, and Peru). In short, it is my belief that the California evidence holds important consequences for our understanding of the origins and developments of complex astronomies among early peoples. In time the data will become especially relevant in formulating interpretative analogs and testable hypotheses. Doubtless it is already time for us to begin thinking of these human achievements as an intellectual continuum. The stimulus was more our ancestors' attempts to understand and cope with their total environment than solely a need to control crops or flocks.

Since this paper is the first of its kind to survey native California knowledge in astronomy, it is important to explain the two major problems involved in its preparation. The first of these is the nature of the ethnographic data. Most nineteenth and early twentieth century ethnographers lacked special training in astronomy and, handicapped further by the assumption of unsophisticated astronomy among California Indians, they usually failed to record much more than general descriptions. The results are gaps in information and apparent misunderstandings or errors. In addition, their work was undertaken at a time in which much of this knowledge had already ceased to exist. Most informants either had little or no first-hand knowledge of the subject (intensified by its restricted use in native times). When informants were knowledgeable, they often were reluctant to share what had been traditionally kept secret. Therefore, contradictions, inconsistencies, and lacunae are found in early ethnographic records.

In generalizing the data for California as a whole we encounter a second problem. We place, perhaps, too great an emphasis on cross-cultural patterns, in contrast to an exhaustive literature search and concentrated analysis on individual cultural achievements in astronomy. My effort has been to strike a balance, I hope with an awareness that not every aspect of astronomical behavior presented can be strictly applied to each of California's native peoples. Certainly the beliefs and practices in astronomy varied from group to group, but for the most part the patterns presented were widely shared.

It is important that the reader understand to some extent the nature of cultural variability in California. Before describing native California astronomy, it is well to begin by describing the setting.

The Setting

California today is a land of contrast, both environmentally and demographically. Such contrasts were also present in pre-contact times. Population estimates on the eve of Hispanic settlement (A.D. 1769) range from 150,000 to as much as twice that number, ranking the area as the most populated in the United States and Canada (Cook 1978). As many as 75 or more mutually unintelligible tongues were spoken. They comprised some 300 different dialects. California was, in fact, one of the most linguistically diverse regions in the world (Shipley 1978).

These various peoples occupied a land of considerable contrast, ranging from dense redwood forests in the northwest to arid deserts in the southeast. Abundant plant and animal resources were efficiently exploited so that large human populations were supported in many areas. Cultural responses varied with the diversity and abundance of natural resources, except in southeastern California where some agriculture was practiced among the Mohave, Yuma, Kamia, and others.

But while cultural responses to land and sea varied much from region to region, the pattern of socio-political organization was widespread and uniquely Californian. Rather than being grouped in tribes (a cluster of social groups forming a single entity characterized by a common language, territory, cultural heritage, and socio-political structure), the Californians were organized in far smaller, more fragmented entities which anthropologists have given the special name of "tribelets." This organizational pattern was found over much of

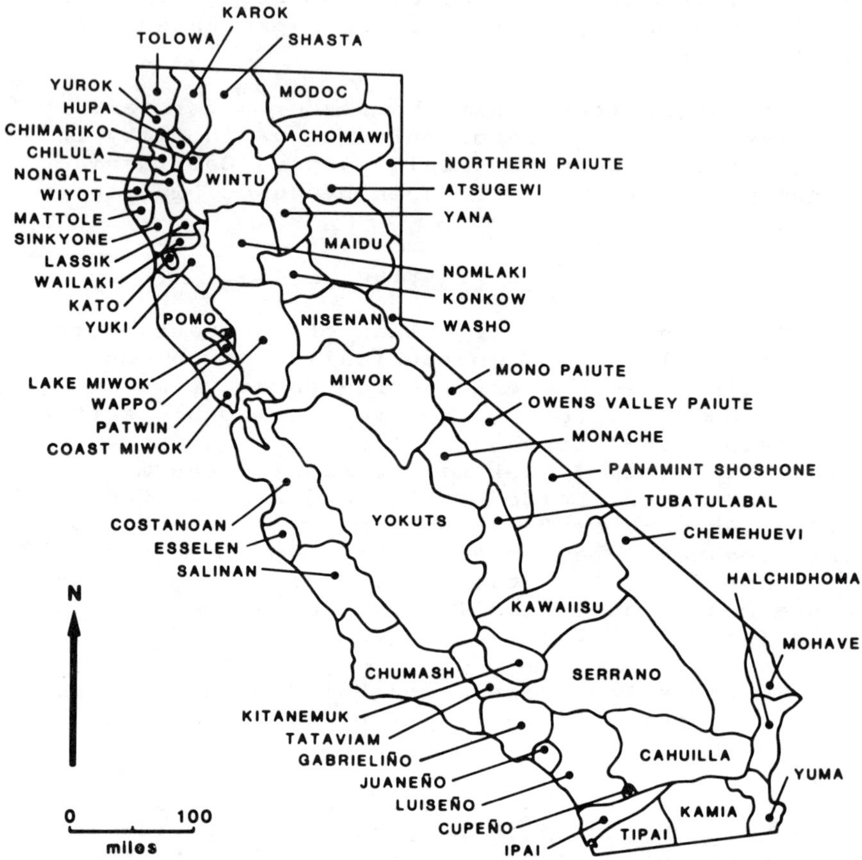

At the time of historic contact with Europeans, Native Californians comprised a linguistically and culturally diverse group of ethnic nationalities, whose territories are mapped here. This map is based on several sources compiled by the author. (Griffith Observatory)

the State, except along the Colorado River and in some parts of the northwest. Tribelets were autonomous, self-governing, and independent socio-political entities. They usually consisted of one or more villages, sometimes as many as a dozen, with the largest serving as a capital of sorts from which the surrounding territory was governed. Leadership was vested in a chief, aided by various officials, who also held social and religious offices.

It is estimated that there were as many as 300 tribelets in California, perhaps corresponding with the various linguistic dialects. For convenience, anthropologists have grouped the tribelets into broader, non-political ethnic nationalities, made up of people who spoke a common language or a collection of closely related languages and shared a similar cultural pattern. Pomo, for example, refers to about 50 or so tribelets, speakers of several related Pomoan languages. I will follow the established practice of using these ethnic nationality terms, although it should be pointed out that astronomical beliefs and practices surely varied among the tribelets within these larger groupings.

Each tribelet was a ranked society, hierarchically structured. Responsibility for maintaining the tribelet politically, economically, and religiously rested with the upper (elite) class. Protection of the tribelet from all potential sources of danger entailed a number of philosophical assumptions about the universe (Bean 1974,1975). It is essential to understand the nature of these philosophical assumptions in order to understand the astronomer elite who act upon them.

The Universe

The most important assumption about the nature of the universe was that it contained power, and the use of power by various beings was the principal causative agent for change. Native cosmologies include not only the creation of the universe, but also the creation of power. Bean (1975:25) points out two principal patterns found in native California:

(1) Power is created from a void in which two forces, usually male and female, come together in a cataclysmic event that forms a creative force; or (2) Power and cre-

> ators appear simultaneously in the universe without explanation, and a creative force begins thereupon to form or alter the world. In both cases, various acts are accomplished through a use of power by a creator or primary creators that leads to a series of creations--among which is man. The outcome of these acts ...is the creation of a hierarchically structured social universe--a cosmological model in which the nature of power is defined and rules are established for interacting with power sources. Although accounts in individual cosmologies may vary considerably from group to group in depicting this universe, throughout California there appears to be at the least, a tripartite division of the universe into upper, middle, and lower worlds.

In form, these worlds were considered flat, circular in shape, and surrounded by either a void or by water. They were placed one above the other and sometimes supported by various beings. Each world contained considerable power, which had been scattered since creation. The lowest world was very dangerous and filled with various powerful, malevolent beings. The middle world was the home of people, and it was also inhabited by various powerful non-mortals. The center of this world was held to be the center of the universe, and each tribelet identified this center with some particular feature within their territory.

Of greatest significance here was the upper world, for it was the home of powerful celestial objects who were the beings responsible for most of the significant events taking place below. Throughout much of California these beings were believed to have ascended into the upper world from the middle one to escape death. They had some of the same characteristics seen in people--will, emotionality, and intelligence. When using their power, the Sky People could be positive, negative, or neutral. Generally, they were considered to be more negative (malevolent) than positive (benevolent) because of their human-like ability to be unpredictable. For this reason, the upper world was conceived as highly variable and dangerous (Bean 1975).

Fortunately, people could also acquire and use power, and since people were near the top of the power hierarchy in the middle world, they could cope, with varying degrees of success, with these celestial beings. In ancient times people had been provided with knowledge (power) and rules for its use. Accordingly, they had a responsibility to interact with these sky beings to maintain a balanced and viable universe. It was therefore up to people to apply their power properly to prevent any particular supernatural from gaining the superiority that would create a cosmic imbalance and death to all (Bean 1975:25-27; Blackburn 1975).

The celestial struggle for power was a continuous process. It pitted the sun, the moon, some planets, significant stars, and constellations against each other. Individually and as "teams," they tried to gain control over the universe. These constant struggles caused changes in the middle world: seasons, weather, animal migrations, and disease. To understand these changes required understanding and interacting with the many beings in the sky who were responsible for them. And this, in turn, required daily observations and ritual activity.

The moon's nightly changes were observed, for example, and every month a New Moon Observance ritual was held in order to help return the moon to its previous state of being. In addition, the two solstices were critical times when Sun, an immensely powerful being, reaches an extreme point in his travels and might decide not to return. Should Sun desire to continue on and not come back, it would mean that cosmic imbalance was inevitable. It was therefore essential for man to predict when Sun was approaching this decision, and to undertake the proper use of power to encourage him to return. Solstice rituals for this purpose were practiced in most of central and southern California (Hudson, Lee, and Hedges 1979).[1]

Among tribelets in central and southern California another dimension of the power struggle was seen as taking place between two teams of Sky People, which met each night to gamble. The Chumash considered Sun as captain of one team, which included Evening Star; the opposing team was led by Polaris and included Morning Star. Moon kept score of the game. Among the Kitanemuk the players were probably the same, while some Yokuts tribelets identified them as Sun and *Tihpiknits*, a feathered supernatural who may have been Polaris. The stakes of the game were the same, regardless of the players--life if one team won, and death if their opponents won. Man's role was to observe, and at the time of the winter solstice when the yearly game had finally ended, it was up to him to determine which team was the winner (Hudson and Underhay 1978 32-33; Hudson and Blackburn 1978; Hudson, Lee, and Hedges 1979; Blackburn 1975).

Other cosmic events, such as an eclipse or comet, required human involvement to maintain a balanced universe. Observation, followed with interpretation, was needed before the power could be properly applied. For the astronomers to cope with such unusual events required knowledge of past events (as explained metaphorically in myths), interpretation (reasoning from assumptions about reality), and expression in ritual behavior (prayers, songs, offerings, ceremonies). It was the elite who accepted these tasks, and it was the astronomers who were required to give them viable expression. For them, astronomy involved a complex understanding of behavior of the various Sky People, among whom Sun was the most important.

Sun

Throughout California, Sun was endowed with vast powers and was seen as a vivifying male force. Among the Chemehuevi in the southeast, he was addressed with an honorific title, and his beard was symbolized by the rays of a sunrise-sunset sky (Laird 1976: 89-90). In the northeast, Sun was in mythic times a great village captain (Voegelin 1942: 143). The Chumash considered this captain as bearded and naked, carrying a fire torch to light and heat the world, and traveling across the sky following a cord stretched around the world (Blackburn 1975; Hudson and Underhay 1978:36-37). The Yokuts believed Sun's east to west journey ended when he reached a lake (Pacific Ocean?) which he entered every night, only to come up again on the other side of the world every morning (Gayton 1948: 88-89). The Atsugewi said he returned to the east via the south side of the earth. As he and Moon passed one another on the other side of the world, the two would decide the weather (Garth 1953:195).

In most, if not all, tribelets, prayers were delivered daily to Sun. In northeastern California people talked to Sun (Voegelin 1942:143); the Pomo asked Sun for power to cure (Loeb 1926:227). If someone was sick from Sun, a doctor could cure him by making an image of Sun, addressing a prayer to that being, and making offerings of heads and feathers. Members of the Pomo secret society were the ones who prayed to Sun when they were doctoring. The Yuki also involved Sun in curing; their shamans made rock paintings and delivered prayers to Sun (Kroeber 1925:198). Prestigious Yokuts shamans were said to have acquired their powers from Sun, whom they addressed with prayers (Gayton 1948:88-89, 229). The Chumash, Kitanemuk, and Gabrieliño also addressed prayers to Sun (Blackburn 1975; Hudson and Underhay 1978:61-72), and one Chumash informant stated that prayers were made daily to the rising sun (Harrington n.d.). Depictions of Sun due to such activity are therefore to be expected, and some examples are evident in rock art in the State.

The critical times for people and the earth were the solstices. Then it became necessary to predict the approach of a possible cosmic imbalance and to perform rituals to prevent it. Among the Chumash, and perhaps their Gabrieliño neighbors, solstice ceremonies were undertaken by the leader of the secret society, who was called "Image of the Sun" or "Splendor of the Sun," and was aided by twelve helpers termed "Rays of the Sun." Their activities involved erecting a "sunstick" oriented with the sun at zenith to "pull" Sun back again, making offerings and prayers at shrines, and performing at least some rock art (Hudson and Underhay 1978:51-73; Blackburn 1975).

Similar beliefs about the solstice appear elsewhere in California, among the Fernandeño, Kitanemuk, Salinan, Costanoan, Tipai, Luiseño, and Costanoan (Hudson, Lee, and Hedges 1979). In central and northern California the winter solstice was connected with general curing ceremonies as well as with displays of supernatural power by shamans (Gayton 1948:267; Driver 1937:86; Spier 1930:240). In much of California the event was also tied into a solar calendar. In a survey by Hud-

The sun seems to be depicted as various rayed disks in California Indian rock art (sources: Heizer and Clewlow, 1973, II--figures 1j, 12d, 19e, 47a, 75a, 138e, 168e, 197h, 199b, 214u, 223b, 253c, 285n, 364a, 364b; Hudson, Lee, and Hedges, 1979--figure 4. Griffith Observatory)

son, Lee, and Hedges (1979) it was found that the solstices were regularly observed throughout most, if not all, of California, and information from these observations being used to regulate a calendar either directly (turning point in the year), or indirectly (by recording the lunation in which the solstice occurred). This calendar, in turn, regulated secular and nonsecular behavior.

The survey also found that two observational techniques were involved: direct observations, in which the sun was watched in relation to the horizon or horizon markers (reported for 16 groups) and indirect observations, in which a beam of sunlight was observed. This beam might strike a mark or series of marks placed on the floor, center post, or wall of a ceremonial structure. Or it might strike a painting or an object placed in a rockshelter. Possibly it was used to cast a shadow (reported for 10 groups). Each of these methods was practiced, and some tribelets apparently used both techniques.

Evidence of solstice observatories in pre-contact California is now emerging, based upon archaeoastronomical studies conducted since 1975. To date, some 10 sites have been reported, and others are being investigated. Summaries of the work of several investigators are reported in Hudson, Lee, and Hedges (1979).

"Direct" observatories involve watching the sun's daily position change on the horizon. The technique is reported for the Wintu, Atsugewi, Achumawi, Shasta, Klamath, Modoc, Maidu, Knokow, Nisenan, Pomo, Chumash, Cupeño, Tipai, and Yuma in the ethnographic literature (Hudson, Lee, and Hedges 1979:Table 1). Direct solar observatories are suspected archaeologically at six California sites: one in Kern County, two in Santa Barbara County, one in Los Angeles County, and two in San Diego County (Hudson, Lee, and Hedges 1979).

The Kern County site contains a rock art motif which may depict the positions of winter and summer solstice sunrise in relation to the horizon, as seen from the site. Robert Schiffman, who tested the prediction by on-site observations, found that solstice positions of the sun agreed with those indicated in the rock painting.

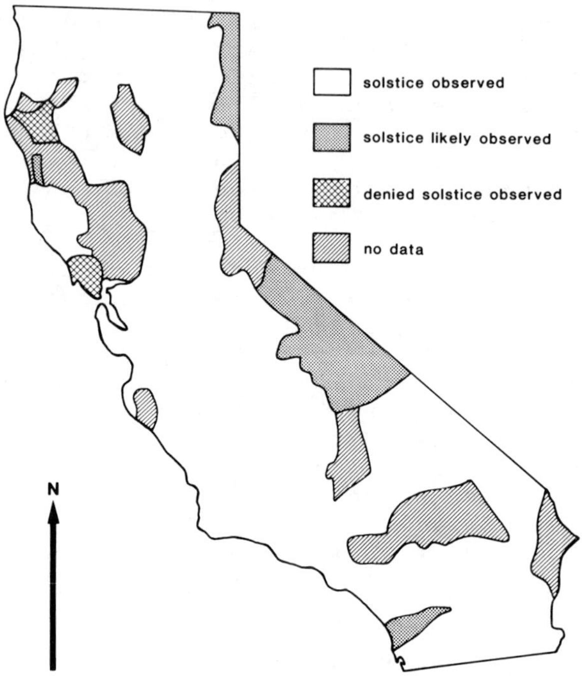
solstice observed
solstice likely observed
denied solstice observed
no data
N

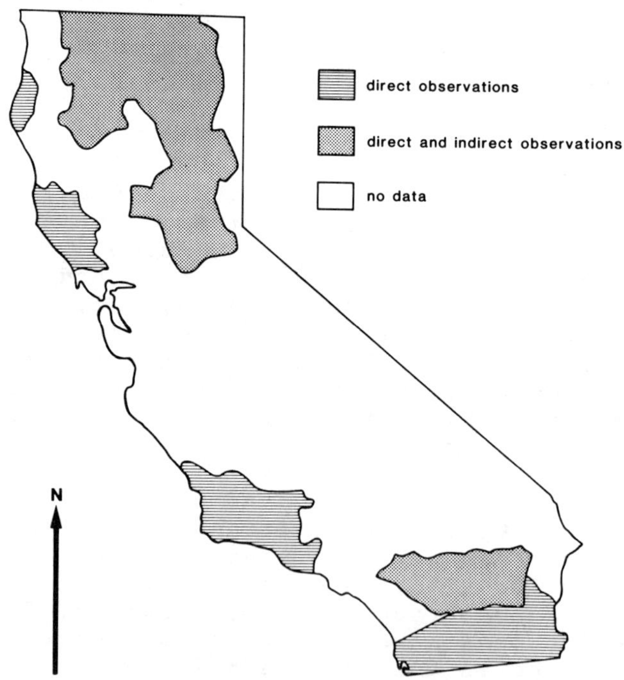
direct observations
direct and indirect observations
no data
N

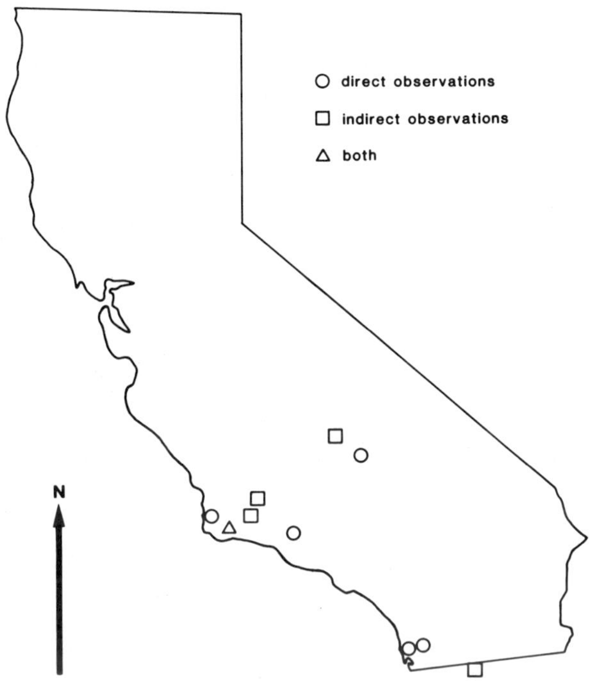

Indians throughout much of California are known to have observed the solstice (top left). The ethnographic reports also permit the mapping of the use of direct and indirect techniques of observation (bottom left). Several archaeological rock art sites where direct or indirect observation may have taken place are now known (above). (after Hudson, Lee, and Hedges, 1979. Griffith Observatory)

In Santa Barbara County, rock art at the Honda Ridge site also indicated a solstice association. Observations by Larry Spanne showed that the point of sunrise was well defined against a fixed horizon. Nearby Window Cave, with petroglyphs including a sun symbol, was also investigated by Spanne. Here he saw that the sun, framed by a small opening in the rear of the cave at winter solstice, sets on the summit of a known Chumash shrine.

Castle Peak, another ethnographically identified shrine, is located in Los Angeles County on the boundary between the Chumash and Fernandeño. On the summit of the peak is a small, natural rock window which is oriented northeast. During the summer solstice John Romani and others observed that the rising sun was tightly framed within the window. Bead offerings found at the site indicate that the shrine was in use between A.D. 1500 and 1850.

San Diego County's two sites are within the territory of the Tipai, and both, examined by Ken Hedges, were found to involve manmade rock alignments, oriented to the position of winter solstice sunrise. One, at Cowles Mountain, consists of a halved circle, the bisecting axis oriented to winter solstice sunrise. Observations by Hedges have confirmed the alignment. The second site, situated on Viejas Mountain, incidentally, is ethnographically described as the place in which the Tipai performed ceremonies at winter solstice to honor the sun.

"Indirect" solar observatories are more difficult to identify, even though the practice is reported for a number of California groups, such as the Shasta, Klamath, Modoc, Achumawi, Atsugewi, Maidu, Konkow, Nisenan, Wintu, and probably the Cahuilla. So far, five sites have been reported which involve the association of rock art and solstice sunlight. Three of these are in Santa Barbara County, one is in Tulare County, and another is just across the border in Baja California.

The three Santa Barbara County sites are within Chumash territory. The first of these is Condor Cave, where Travis Hudson and Ernest Underhay observed a beam of sunlight to enter the cave through a tiny, manmade window at the time of winter solstice sunrise. Rock art is found in the cave. Another rockshelter, not far distant, contains a set of concentric circles painted upon a natural stone column. During winter solstice sunset, Steve Junak observed that light entered the cave and bisected the motif diagonally. The third site, Window Cave, has already been mentioned for its association with a nearby mountain as a direct observatory. Observations by Spanne at the site indicate that it may have also served for indirect observations as well. During winter solstice sunset a ray of sunlight strikes one wall of the shelter as a long, narrow band, directing the viewer's gaze toward a probable sun motif.

At La Rumorosa in northern Baja California, in Tipai territory, a complex pattern of sunlight has been studied by Hedges as it strikes the wall of a painted rockshelter at winter solstice sunrise. The incoming light first strikes a circular design painted in white, and then begins to extend in a long, pointed steak toward a red-horned anthropomorphic figure. The sunlight skips over the figure, which is located in a slight recess, and shines on the wall beyond, forming a sharp point of light which gradually extends back toward the sun and falls across the eyes of the figure. As more light enters the rockshelter, the rest of the figure's body becomes illuminated. This unique play of sunlight is known to occur only within ten days of winter solstice.

The Tulare County site, in Yokuts territory, is a painted rockshelter with a large southeast entrance. One of the more conspicuous polychrome paintings on the ceiling is a large figure which is oriented southeast. The animal appears to be a coyote with what may be a solar symbol in its mouth. The association of Sun and Coyote with the winter solstice among the Chumash has previously been noted, and it occurred to Hudson and Underhay that perhaps the depiction in this cave was likewise associated with the winter solstice. Observations by Underhay of sunrise on the winter solstice indicated that a ray of light entered the shelter and illuminated a small, painted recess in the northwest wall.

Motifs in California rock art which may show that particular sites were connected with direct solar observations have been described by Hudson, Lee, and Hedges (1979: Figs.4,5). These motifs share two features: a wavy-line, zigzag, or simple curve which is horizontally-oriented, and one or more circular figures which may represent the position of the sun. The connection between rock art, the sun, and shamanism has already been noted, and additional support is found in ethnographic descriptions of how the solstices were determined. The Wintu, for example, waited for the sun to oscillate between two points on the horizon for three days and then reverse direction (Voegelin 1942:234). The Pomo proclaimed the solstice when the sun rose over the same spot four times (Loeb 1926:223, 226). One Chumash sunwatcher proclaimed the solstice by observing the sun's changing position in relation to three peaks (Blackburn 1963: 141). There are other descriptions as well, all of which strongly support ethnographic statements that the sun and the solstices were important to native Californians.

Little is known about California Indian observation of the equinoxes. Cope (1919:121) surveyed the question of whether the equinoxes were known in native North America. She concluded that everywhere the event was of little importance, although known among the Tewa in the Southwest and the Nootka hunter-gatherers of the Pacific Northwest Coast. In California, the Chumash appear to have been aware of the equinox, noting it in terms of calendar days (Hudson and Underhay 1978:126). Goldschmidt (1951:388-389) reported that the Nomlaki observed the equinox, although there is some confusion with his informant over whether or not the solstices were also known and observed.

Among most northern groups, for example, the Wappo (Driver 1936:195), Moon was symbolized as a "night sun"; the Wintu referred to the moon as the "silvery underside of Sun's belly" (DuBois 1935:75). The Shasta singled out the moon (new?) nearest either solstice point as "Big Moon" (Holt 1946:342).

Generally, throughout California, Moon was Male; however it was identified as female by some tribelets, including some Pomo ones (Gifford and Kroeber 1937:148) and the Chumash. And apparently the Chumash saw Moon as a supernatural being equated with *datura*, a powerful hallucinogen which could provide power to cure, to foresee the future, or to acquire additional power (Blackburn 1975; Hudson and Underhay 1978:77). All groups believed Moon had the ability to regulate both health and a host of natural cycles. The Luiseño, for example, believed that Moon watched over the people to regulate all things, and that people were strong or weak depending upon whether Moon was waxing or waning (DuBois 1908:164). The Chumash considered Moon's powers to be intermediate between those of Sun and Earth. She affected all living things, controlling the menses of women, the tides, and the cycles of other earthly creatures (Blackburn 1975; Hudson and Underhay 1978:75-76). If a Pomo shaman had cause to suspect that Moon was creating sickness, he countered the effects by making an image of Moon and delivering offerings and prayers (Loeb 1926:227-228).

Throughout California the moon's phases were observed and named. Equally widespread was the belief that Moon died monthly and was reborn. The most important lunar phase was the new moon, a time when intense ritual activity was undertaken to help Moon. The Luiseño, for example, said that they knew on what evening the moon would be reborn, and they conducted their New Moon Observance accordingly (Harrington 1933:179-180).

Typically, the New Moon ceremony, over the entire State of California, consisted of shouting and prayers for good health and long life, and everyone came out to take part. In some regions races were conducted (Driver 1937:88; Drucker 1937a:26; Aginsky 1943:427; Barnett 1937:193; Kelly 1932: 152-154; Gayton 1948:89, 229; Laird 1976:90-91; Harrington 1933:179-180, 1942:42). The Pomo considered the new moon an unlucky time because of its association with menses; men could not hunt at this time (Loeb 1926:227-228). The Atsugewi considered the new moon to represent strength and life (Garth 1953:195), and prayers to the moon at this time expressed a common desire for good health and long life. Wintu head men also addressed prayers and offerings of food to Moon at this time (Voegelin 1942:145). In northwestern California, speeches and prayers were made; the Yuki and Sinkyone used kin terms, and older individuals asked Moon to give them new life (Driver 1939:342-343). The Luiseño believed that just as the moon died and was

SISKIYOU

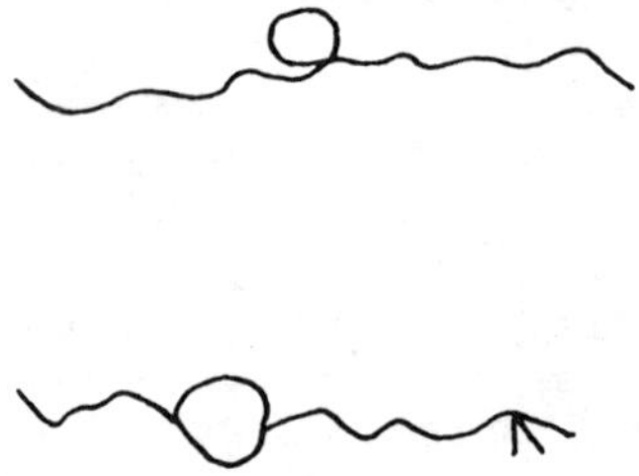

LASSEN

SAN BERNARDINO

MONO

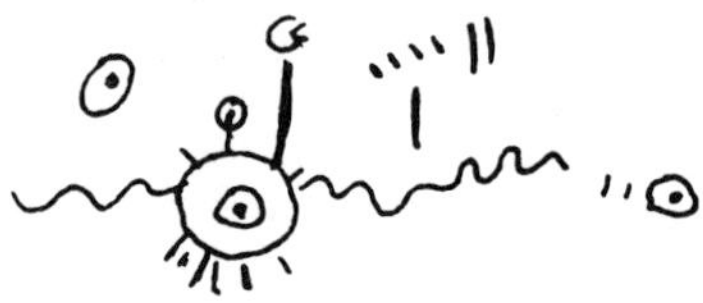

MODESTO

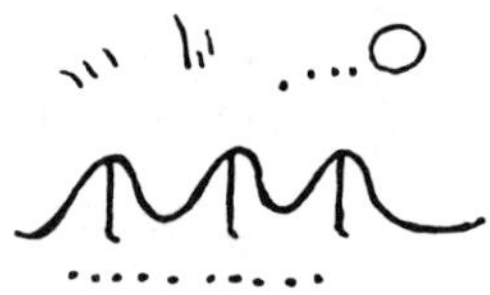

MODESTO

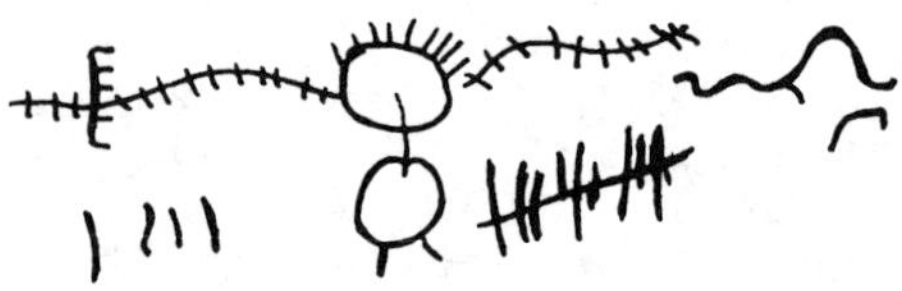

MODESTO

SAN BERNARDINO

MODESTO

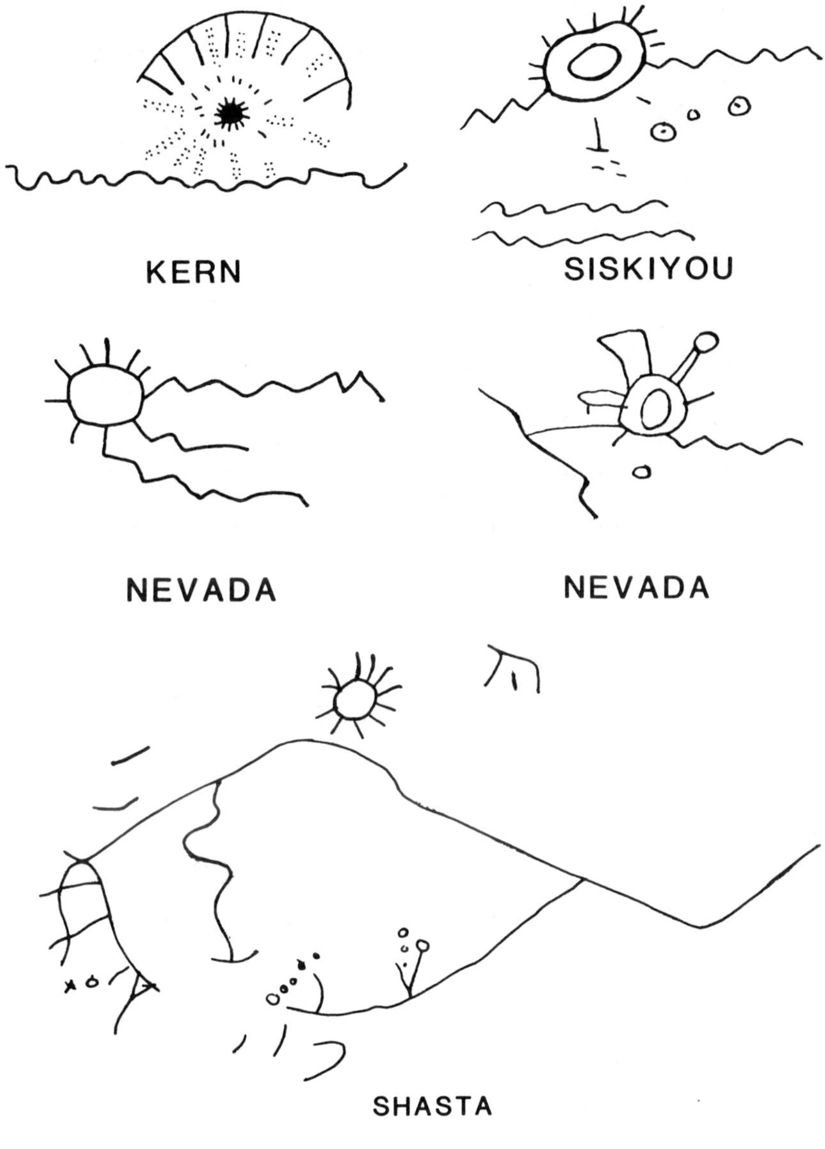

A number of possible depictions of the sun on the horizon--possibly references to solstice observations--are known from California rock art. (after Heizer and Clewlow, 1973, II--figures 105d, 105e, 130j, 151b, 153b, 154b, 155b; Smith and Turner, 1975--50/H23, 78/H60; Hudson, Lee, and Hedges, 1979--figure 6a; Heizer and Clewlow, 1973, II--figures 84a, 176e, 176g, 247; Steward, 1979--figure 3. Griffith Observatory)

born again, so also do we die and are born again (Harrington 1933:179-180). This belief was no doubt connected with the ritual death of shamans and their revitalization with supernatural powers. The Chumash also addressed the new moon with prayers, and they extended the arms outward while praying (Harrington n.d.).

The reappearance of the moon as a brilliant crescent was believed by people in all of California to indicate approaching climatic conditions. Observers watched the position of the "horns" or "points." If the crescent was horizontal (the points side by side) it was a sign of potential rain. If it was vertical (one point above the other), it was a sign of drought and equated with death. The distribution of this belief is correlated with the locations of California's regions with the Mediterranean climate of wet winters and dry summers (Drucker 1937a:26; Driver 1937:88; Aginsky 1943:427; Sapier and Spier 1943:283; Steward 1941:267, 270; Harrington 1942:42; Zigmond 1977:60; Hudson and Underhay 1978:76). Considerable variability in what native Californians considered the horn positions to indicate is found in northwestern California (Driver 1939:342-343), which does not have a Mediterranean climate.

The moon was observed daily, but we do not know if these observations were made from a special location, as was the case for the sun. Perhaps both celestial objects were observed from the same site. Throughout California the day of new moon was used to begin the month, and various counts were maintained over a period of time. The Surprise Valley Paiute, for example, recorded the occurrences of new moons by making extra long scratches on a stick (Kelly 1932:152-154). Probably the new moon's association with the beginning of a new life contributed to its use in determining the monthly beginnings in the calendar.

It is not surprising to find the crescent to be the most common lunar symbol in California art. The Chumash, for example, painted this crescent form in red (with horns side by side to indicate rain) on the solstice sunstick mentioned earlier (Hudson and Underhay 1978:77), and there is an interesting description of a woven belt used on dance skirts in certain ceremonies which had abalone shell ornaments in the shape of small stars and half moons (Hudson 1979:26). In Ipai ground paintings the lunar symbols used were both the new moon crescent and the simple circle full moon (Waterman 1910:301, 352-353, Pl. 20; Kroeber 1925:662-663, fig. 56f). Circular motifs and crescent are common elements in rock art, and it is quite likely that some of these crescent forms represent the moon.

Eclipses

Everywhere in California both solar and lunar eclipses were known, and these phenomena were considered ill omens. Explanations varied from group to group, but in northern and central California it was generally believed that some animal among the Sky People was eating the sun or moon. In the northwest this animal was either a condor, bluejay, coyote, dog, bear, or raccoon (Barnett 1937:193; Driver 1939:343); in the northeast, as well as among several central and eastern groups (Pomo, Nomlaki, Wintu, Yana, and Nevada Shoshone) the animal was a bear or a dog (Loeb 1926:228; Goldschmidt 1951:388; Spier 1930:221; Voegelin 1942:146; Steward 1941:267; DuBois 1935:75; Sapier and Spier 1943:283).

Explanations for eclipses were treated mythologically. The Pomo believed the Milky Way to be Bear's path, and one myth relates that Bear once met Sun on this path and caused a fight leading to an eclipse. Later, Bear had a fight with Sun's Sister, Moon (Loeb 1926:228-229). For the Paiute an eclipse meant that Sun was dying (Kelly 1932:199-200).

Some Monache and Yokuts groups believed the animal encountered by Sun or Moon was a condor, coyote, or waterbugs (Driver 1937:88; Gayton 1948:267). A Yokuts myth recorded by Latta (1936:137-138) contains an obvious description of a solar eclipse. The story relates how Coyote, desirous of death for all, sought out Sun. When he located the trail on which Sun was traveling, he blocked the path, but Sun made it so hot for him that he "had to coil his tail under him, all but the tip." (The coiled body may have symbolized the black disc moving across the face of the sun.) This narrative again illustrates the celestial power struggle mentioned earlier.

The moon's distinctive phases permit association of some rock art symbols with the moon. In fact, there is no other common and conspicuous crescent in the sky. (after Heizer and Clewlow, 1973, II--figures 27a, 38e, 76h, 86, 88d, 126d, 140c, 187a, 243i, 305e. Griffith Observatory)

The Chumash and some of their Yokuts neighbors believed that a giant bird, such as an eagle or condor, was responsible for an eclipse by hiding Sun or Moon with its wings (Driver 1937:88; Aginsky 1943:427; Blackburn 1975; Hudson and Underhay 1978:71-72). The Yokuts condor dancer's headdress consists of crescent horns (Driver 1937:105, 143-144), and it is conceivable that this dancer's costume, especially the headdress, symbolized an eclipse in some way.

The Miwok explained eclipses as a power struggle between Sun and Moon: The beings were fighting, and if Sun were to win, the world would be consumed by fire. Understandably, the shamans helped Moon to win by making offerings (Aginsky 1943:427).

In southern California there was the widespread belief that Sun or Moon was being eaten by spirits. Shamans would watch to see who would die and then attempt to call back their spirits (Drucker 1937a:26). The Luiseño associated an eclipse with sickness, an idea with mythological basis in the account of *Wiyot* being slowly poisoned and dying, only later to be resurrected as Moon (DuBois 1908:164; Harrington 1933:179).

Since an eclipse represented a definite imbalance in the universe with potential death to all, various forms of ritual activity took place at these times. Throughout California, everyone shouted during an eclipse in hopes that the noise would drive off the particular animal devouring or hiding the object, and thereby help Sun or Moon come back to life (Spier 1930:221; DuBois 1935:75; Barnett 1937:193; Driver 1939:343; Loeb 1926:228; Gayton 1948:267; Garth 1953:195; Voegelin 1942:146; Goldschmidt 1951:388). Unfortunately, the more formalized ritual activity which took place is little described. In northwestern California, Yurok men would go into the mountains to make medicines (Driver 1939:343). In the central part of the State some groups believed that a condor was eating Sun, so an animal would be killed and its carcass left out to attract the animal away from Sun (Gayton 1948:162). Among the Kawaiisu, an eclipse of the sun or moon was considered a sickness, and an old man would undertake to cure Sun or Moon by singing and dancing; "he had 'dreamed' the [curing] song" (Zigmond 1977:89).

KERN

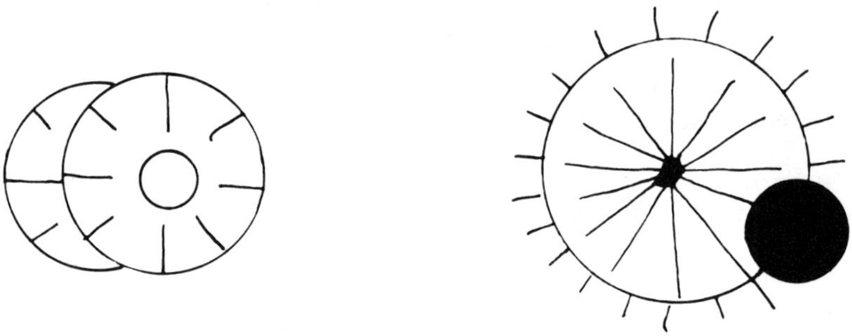

TULARE

Eclipses also may have found their way into rock art. Although these symbols may be interpreted in other ways, there are aspects of all of them that suggest the idea of a solar eclipse, in which the dark disk of the moon obscures the brilliant disk of the sun. (after Heizer and Clewlow, 1973, II--figures 86h, 371b, 375b. Griffith Observatory)

Other formalized eclipse ceremonies are indicated for southern Californians, particularly the Luiseño, Kamia, and Tipai, and perhaps their neighbors. The Luiseño gathered to sing songs to help Moon get well (Harrington 1933:179). Songs and dances were performed by Kamia men, whose purpose it was to prevent Moon's total extinction (Gifford 1931:66). Among the Tipai, as soon as the eclipse started the chief would begin the ceremony by singing "the song of the people" (Spier 1923:326). It is not known whether the timing of these various ceremonies and the necessary preparation (gathering people together or killing an animal to lure condor away) represent an ability to predict an eclipse.

It is probable that eclipses were portrayed in art, but no study of such activity has yet been fully made. Hudson and Underhay (1978:71) have noted possible solar eclipse motifs at two rock art sites in Chumash territory, and examples of others are suggested in California rock art.

The Planets

There is a dearth of information about the planets. In his general summary on California Indians, Kroeber (1925:682, 874-875) suspected that for most tribelets, excepting perhaps the sky-oriented Luiseño, these objects would have been too difficult to trouble with. Recently, Hudson and Underhay (1978:79-86) have assembled evidence that Venus was observed in some regions of central and southern California; quite possibly among the Chumash, Mars, Jupiter, and Saturn were observed as well.

Throughout California, Venus was named and identified separately as Morning and Evening stars by groups such as the Nomlaki (Goldschmidt 1951:389), Wintu (DuBois 1935:75), Nisenan (Beals 1933:357), Pomo (Gifford and Kroeber 1937:190; Loeb 1926:228), Wappo (Driver 1936:195), Luiseño (DuBois 1908:162; Harrington 1933:200-201), Chumash (Hudson and Underhay 1978:80-81), Chemehuevi (Laird 1976:91), Yokuts (Gayton 1948:89, 156, 229), and Monache (Gayton 1948:267). In short, it is evident from this list that the planet Venus was important.

Perhaps Venus, like the stars, was believed by native Californians to be one of the First People who had ascended to the sky. If so, it would be important to know which of the First People this planet represented, and whether it was seen as a single being changing positions in the sky or as separate beings. Some groups recognized Venus separately as Morning and Evening stars. The Pomo, for example, considered Morning Star as "Morning Eye Fire," and Evening Star as her younger sister (Loeb 1926:228). The Chumash conceptualized Venus as separate beings also. Morning Star represented two benevolent beings (Two Thunders) associated with lightning, thunder, and rain; Evening Star was a malevolent being (Golden Eagle) associated with death and afterlife (Hudson and Underhay 1978: 80-90). Gayton (1948:89, 156, 229, 267), however, reports that the Yokuts and Monache considered Morning and Evening stars to be the same object. It simply switched places in the sky. The neighboring Kawaiisu perhaps considered them as two separate beings. Coyote, it was said, threw Evening and Morning Stars into the sky while he was playing ball, and that, it is said, is why these stars can be seen bobbing up and down behind the mountains (Zigmond 1977:92).

Some native Californians addressed prayers to Morning Star. Loeb (1926:228) reported that the Pomo did so because it aided in all events. Gayton (1948:89, 156, 229, 267) stated that the Yokuts and Monache delivered prayers to Venus. All men of character arose early enough to greet Morning Star and request its help to solicit aid from Sun and Moon. This was probably because Venus rose in the east, where Sun and Moon emerged from their houses, and therefore it was believed to have some connection with these beings. A Chumash informant also noted that the people would rise up early to pray to Morning Star, extending their arms outward in the same manner used to address the stars (Harrington n.d.).

We do not know at present whether other planets were similarly recognized. Hudson and Underhay (1978:90-96) suggest that the Chumash observed Mars and equated it with a being who was called upon to locate celestial objects which were missing (invisible due to conjunction) in the upper world. Ritual equivalents of this being can be found among the Yokuts (Driver 1937:105, 143-144), and Pomo (Loeb 1926:384-385), and possibly also among other central California groups such as the Maidu, Patwin, and Miwok (Gifford 1955:287-288).

Perhaps planetary observations were conducted from locations of ritual significance, as was the case with the sun. There is a hint that some California groups may have determined in advance when Venus would be in inferior conjunction (passing between earth and sun). The Chumash and probably the Yokuts and Monache associated Evening Star with death. The irregular schedule of an important ritual--the Mourning Ceremony and burning the things of the dead--every two or three years could be explained as coinciding when inferior conjunction was approaching (Hudson and Underhay 1978:94). Latta (1949:232, 238) described one Yokuts Mourning Ceremony in which the burning of the belongings of the dead coincided with the appearance of the Evening Star; a few days later the Morning Star was visible. While no mention is made by Davis (1919:9,21, 23, 25, 33) to the appearance of the Evening Star, he did note its importance in the Ipai-Tipai Mourning Ceremony. As with other Californians, the Ceremony lasted about a week, the timing apparently being irregular to non-Indians, or as Davis (1919:25) noted "...the Indians jealously guarded all knowledge of the time when it is to occur." Davis attended one ceremony performed at Weeapipe rancheria from August 24th to 30th, 1908, and noted that an image house was constructed in the west, while the pit in which they were to be burned was located in the east. The ceremonial direction was thus west to east. During the final evening:

> The dance was resumed and continued without further intermission until the morning star, *Cuanyamasahp*, rose above the rim of the mountains in the east. Thereafter it was closely watched during the dance for an hour and a half, until the surrounding mountains were dimly visible against the first gray light of approaching dawn. It was then 4:15 o'clock in the morning of August 30.... With El Nun [the captain who had begun the ceremony] in the lead, chanting the final canto..., the last procession formed and marched with slow and stately steps directly toward the morning star... [Davis 1919:23].

The image house in the west was then set on fire,

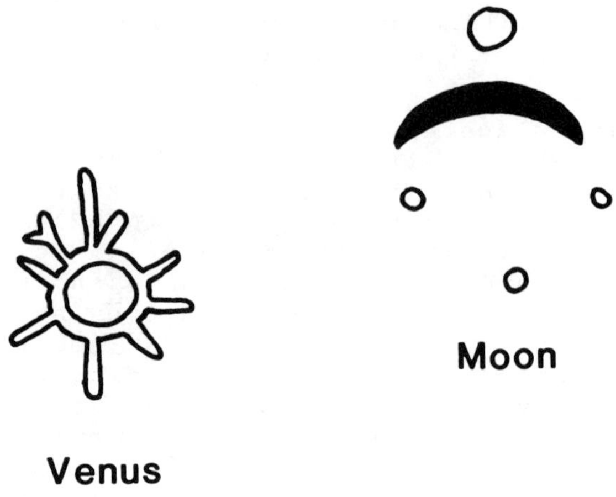

SANTA BARBARA

The brilliance of Venus may have prompted Chumash skywatchers to single it out as a prominent, flamboyant image in the vicinity of the crescent moon. (after Hudson, no date. Griffith Observatory)

---

followed by the burning of the images in the pit in the east.

The two dates given by Davis (1919) for Tipai-Ipai Mourning Ceremonies are the first week of May, 1904, and August 24-30, 1908. He notes that for both, the burning ceremony occurred with the appearance of Venus as Morning Star. The Evening Star was, however, not visible prior to either ceremony. This suggests several possibilities:

1. The conjunction of Venus was not used to time this ceremony,

2. Another planet had been visible in the west and was "confused" with Venus as Evening Star or was treated as having similar supernatural powers, or

3. By late historic times (twentieth century), astronomical ability for precise dating of conjunctions had begun to decline among the elite.

Unfortunately, other ethnographers describing the widespread Mourning Ceremony have failed to note whether either aspect of Venus was visible. Strong (1929:123) mentions that the Cahuilla performed the burning when Orion was directly overhead.

Little can be said about the planets as symbols in art. Hudson and Underhay (1978:89-90) have speculated that an eagle perhaps symbolized Evening Star in Chumash rock art.

Meteors and Comets

A Yokuts informant stated that the doctors knew something about shooting stars, but the "plain people wouldn't know" (Gayton 1948:162). That "something" may refer to various sorts of power attributed to these objects. Unfortunately, data on this are lacking.

Throughout most of California meteors were considered an ill omen, usually associated with sickness or death (Driver 1937:87; Harrington 1942:30; Voegelin 1942:146). The Kawaiisu felt so for a falling star which starts high and falls to the horizon (Zigmond 1977:61). In central and southern California some groups viewed meteors with more indifference, explaining them as the soul's journey on its way to the afterworld (Drucker 1937a:26; Harrington 1942:30; Hudson and Underhay 1978:96-97). The northern California Wintu considered meteors to be the spirits of shamans who had died and were traveling to the afterlife (DuBois 1935:75).

In southern California these objects were considered to be the feces of stars. A single, bright meteor was identified with a cannibal spirit, called *Takwich*, by the Serrano, Cahuilla, Cupeño, Luiseño, and Ipai (Hooper 1920:364-365; Drucker 1937a:26). This spirit lived in a mountain home and would travel about at night looking for wandering souls to eat. Harrington (1933:180-181) stated however, that *Takwich* was not a meteor, but rather was identified with ball lightning. One Cahuilla shaman reportedly received power from *Takwich* as his guardian spirit (Hooper 1920:364-365).

There are other explanations for shooting stars. Some Pomo groups considered them to be fire

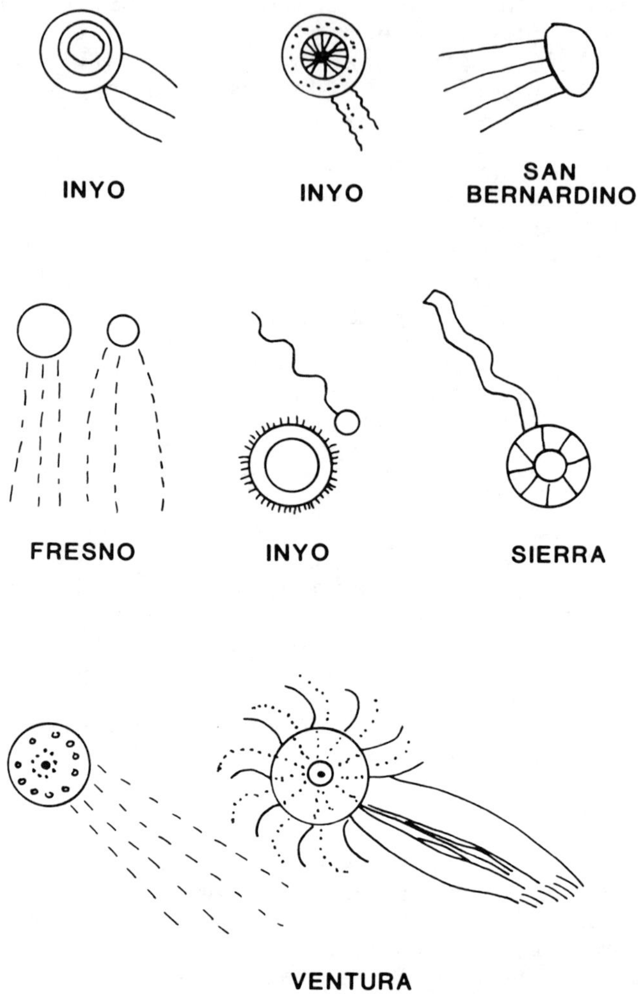

Circular emblems equipped with tails may represent comets, which, when they appear, can look something like these rock art symbols. Bright meteors--or fireballs--also might have been depicted in a similar way. (after Bock, 1974--BC 115; Hudson and Underhay, 1978--figure 14; Heizer and Clewlow, 1973, II--figures 8a, 31g, 45g, 214e, 285i. Griffith Observatory)

dropping from heaven (Loeb 1926:229). Among some eastern California tribelets they were thought to be stars which were changing places in the sky (Driver 1937:87). Harrington (1933:180-181) reported the Luiseño belief that meteors were stars which made a sudden move.

Community rituals probably were not involved with these objects, but a particular meteor shower in the constellation of Gemini may have had important death symbolism to the Chumash due to its occurrence shortly before the winter solstice (Hudson and Underhay 1978:96-97).

Little is known of native California beliefs concerning comets. Harrington (1942:30) reported that the Costanoan considered comets ill omens. The same may be true for the Chumash (Hudson and Underhay 1978:99). The Luiseño also considered comets to be omens, and the object itself was thought to be a former chief who returned with a headband tied around his head (Harrington 1933:201).

Depictions of meteors have yet to be identified in California rock art, but what looks like a pair of comets has been found at one Chumash site; their dissimilarity has led Hudson and Underhay (1978:99) to infer that they represent the same object changing appearance over a span of perhaps several weeks. Overpainting at this site suggests that shaman artists engaged in ritual activity for the entire period in which the object was visible. There are perhaps other depictions of comets in the California rock art.

Stars

In his general synthesis on California Indians Kroeber (1925:874-875) listed the Pleiades as the most mentioned constellation by informants; Orion's Belt was next in frequency, followed by Ursa Major. In addition, southern Californians recognized a half dozen or so constellations and first magnitude stars.

Ethnographic data collected since Kroeber's study indicates, however, a far greater awareness of the stars among native Californians than had been thought. In northern and central California, Kelly (1932:154) reported that the Paiute knew a number of stars and constellations: Barnett (1937: 176) also noted this for the Tolowa, as did Driver (1936:195) for the Wappo. Beals (1933:357) and Garth (1953:195) report that knowledge of the stars among the Nisenan and Atsugewi was limited, a conclusion, I suspect, that was probably based more on an absence of data. In southern California, Laird (1976:91-93) and Gifford (1931:66-70) have noted extensive starlore among the groups they studied, the Chemehuevi and Kamia, respectively. It perhaps can be inferred also for the Kawaiisu, a people who believed that it was an ill omen to fall short when counting stars (Zigmond 1977:61).

Throughout California there were various explanations for the stars, and metaphorical accounts of these objects were perpetuated in mythology. The stars and planets were generally considered to be the First People who had, prior to the creation of man, ascended into the upper world to escape death. A Gabrieliño informant expressed it this way:

> Those who did not want to die escaped to the sky, and that is why all the stars have names. While all that remained on earth are mortals, those who went into the sky always live [Harrington n.d.].

Stars were identified with various animals, plants, and objects; common designations were Coyote star, Buzzard star, and so on. For all of these, there were probably myths which connected them with specific places in the middle world where they resided before ascent. For instance, the Yokuts identified one such place, a rockpile, as the footprints of the Pleiades (Gayton 1948:230).

There is a definite need for researchers to examine myths for astronomical data. There is considerable evidence to correlate many mythological beings with celestial objects, and to determine if the myths have any related ritual behavior. A Chumash myth about Seven Boys has been found to be connected with a ceremony and with the stars of Ursa Major (Blackburn 1975; Hudson and Underhay 1978: 105-106). Songs to stars have also been widely reported. The Luiseño had various star songs, some of which named the stars associated with the seasons (DuBois 1908:162-164). The Nisenan also sang songs to the stars (Beals 1933:357). The importance of this comes out in Kroeber's (1925:682) statement:

> In all southern California constellations are named in ritual, and particularly in song, much more frequently than in the northern part of the State, and play a more important part even than in the ceremonies of the Southwest.

California social organization is also reflected in mythology and in turn perhaps with celestial objects. Hudson and Blackburn (1978:243-244) reported finding in central and southern California a strong correlation between the identities of various clan and moiety organizations with the organization of these various beings in the upper world. It is suggested that the particular being with which each clan was identified also had a celestial representation, as for example a Buzzard Clan with a Buzzard Star, or a Sun Clan with Sun. We do know that the Luiseño associated their Coyote Clan with the North Star (Dougan 1964:142, 144) and that these people also had a Pleiades Clan (Strong 1929: 277). This is discussed more fully later, but the point is that a considerable amount of data on California Indian astronomy can be mined from mythology and social organization.

The widespread pattern of equating stellar beings with various mythological plant and animal beings has one possible exception in California. DuBois (1908:162-164) and Harrington (1933:200-201) reported that the Luiseño identified various bright stars and constellations as star chiefs. Less brilliant stars were considered either elements of the chief's body (a hand is typically mentioned), his costume (a headdress), or his own people (followers who stayed with him in his travels across the heavens). While both of these ethnographers have singled the Luiseño out as having a perspective of the sky different from the rest of native Californians, I strongly suspect that these various star chiefs were synonymous with the First People themselves. For example, the star chief identified with Altair was described by the Luiseño as the Buzzard Chief, an identification also shared by their neighbors.

Unquestionably all native California tribelets recognized and named patterns in the stars they saw, patterns which, more often than not, cross-cut ours or represent what we might consider asterisms. Some of these patterns will be described here, arranged by our classification, rather than theirs. Depictions of various constellations in cult objects, ground paintings, and rock art are known, as perhaps reflected in some of the rock art depictions or native drawings.

Aldebaran

The identification of this star with Coyote is mentioned in various southern California myths and statements, including the Luiseño (DuBois 1908:162-164), Gabrieliño (Harrington n.d.), Chumash (Hudson and Underhay 1978:102-103), Kitanemuk (Harrington 1942:29-30), and Cahuilla (Hooper 1920:365-366). Coyote/Aldebaran should not be confused with Sky Coyote, a different being associated with Polaris. The Chemehuevi considered Aldebaran the leader of the hunters identified with Orion (Laird 1976:91-93). The star was also represented in Ipai ground paintings as "Coyote" (Waterman 1910:Pl. 25; Kroeber 1925:662-663).

Altair

Southern Californians associated this star with Buzzard; it was identified as such by the Kamia (Gifford 1931:68-69), Luiseño (DuBois 1908:162-164), and Tipai (Spier 1923:319-320, 358). A faint nearby star was Buzzard's headdress.

In central California the buzzard symbolism was also present among the Yokuts (Driver 1937:87) and Gabrieliño (Harrington 1933:186, 246). The Chumash and Kitanemuk may have associated Altair with the buzzard also; current data indicate that this star was chief over the Land-of-the-Dead (Hudson and Underhay 1978:121-234).

Antares

This star was identified by the Luiseño as a star chief, but the mythological being is not yet known. His raised right hand was marked by the star Arcturus and signaled his coming (DuBois 1908:162-164).

Cassiopeia

Among the Chumash this constellation was identified with the Land-of-the-Widows. Ceremonies at the time of the winter solstice honored these stars (Hudson and Underhay 1978:119-120).

Capella

Gifford (1931:70) reported that this star was recognized by the Kamia, but its mythological identity is unknown.

Draco

These stars were probably identified by the Chumash as the seven mothers of the Pleiades (girls) and Ursa Major (boys). They were mentioned in myths (Hudson and Underhay 1978:113).

Fomalhaut

This star was identified by the Luiseño as one of their star chiefs (DuBois 1908:162-164).

Gemini

In northeastern California, these stars were identified as twins, a boy and girl. Their bow was a bent arc of six stars which rose before them. Another star following them was Rabbit (Spier 1930: 221). The Chumash recognized Castor and Pollux probably as two sisters, cousins of Sun (Hudson and Underhay 1978:106).

Leo

Either this constellation or Scorpius was stated by the ethnographer as being recognized by the Kamia as *Chiyuk*, a supernatural who rose when Altair set. Gifford (1931:68-69) equated *Chiyuk* with Tipai *Ciluk*, noting in a sketch of the constellation that many very faint stars represented foxes. Spier (1923:319-320, 358) identified *Ciluk* as an arc of stars associated with three large ones. Their mythological association has yet to be worked out.

Lepus

These stars, together with Sirius in Canis Major, formed a pattern named *Mech* by the Chumash. Their connection with mythology and ritual are unknown (Hudson and Underhay 1978:112-113).

Milky Way

Although not a star nor constellation, the Milky Way was considered a single entity by native Californians. Often it was conceived as a ghost's

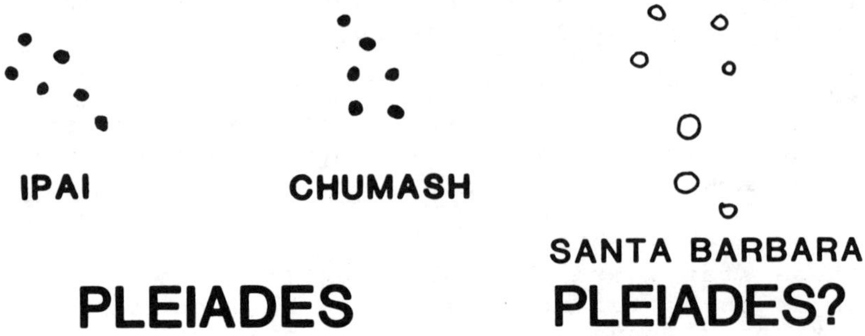
IPAI
CHUMASH
SANTA BARBARA
PLEIADES
PLEIADES?

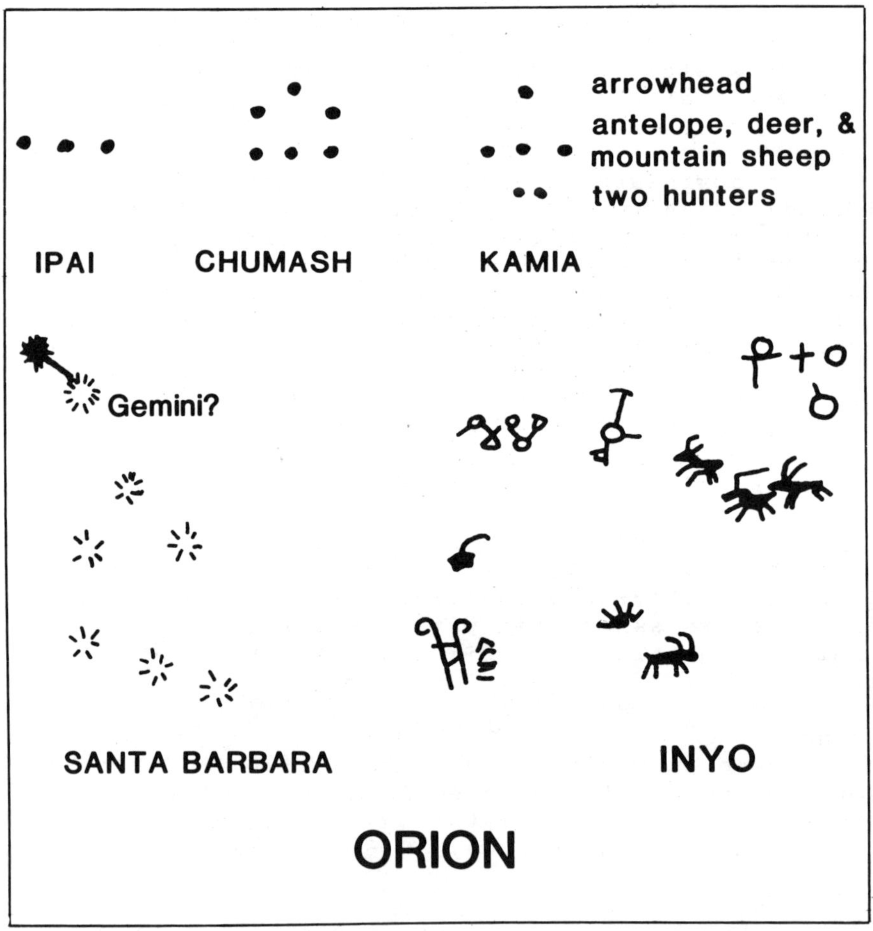
arrowhead
antelope, deer, & mountain sheep
two hunters
IPAI
CHUMASH
KAMIA
Gemini?
SANTA BARBARA
INYO
ORION

IPAI

CHUMASH

ALTAIR

(wolf)
IPAI

(wolf's heart)
LUISEÑO

POLAR STARS

foxes

(CHIYUK)
KAMIA

LEO OR SCORPIUS

Some patterns of stars are so naturally conspicuous that close resemblances between them and rock art designs can be seen. These include the Pleiades, Orion (especially his belt), and Altair and the two stars (β and γ Aquilae). Star depictions identified with the celestial Wolf may represent Polaris. (after Waterman, 1910--plates 24 and 25; Gifford, 1931--68 and 69; DuBois, 1908--figure 3; Hudson and Underhay, 1978--figure 18; Heizer and Clewlow, 1973, II--figure 60c; Grant, 1965--plate 21; Hudson, no date. Griffith Observatory) road, the path taken by the dead to reach the next world. This concept, with multiple symbolic associations, was present among the Luiseño (DuBois 1908:162-164), Chemehuevi (Laird 1976:113-119), with less elaboration among various central California groups, such as the Paiute (Kelly 1932:154) and Nevada Shoshone (Steward 1941:267), and throughout northeastern (Driver 1936:195; Garth 1953:195; Voegelin 1942:146-147) and northwestern (Driver 1939:344) California.

In much of central California there were other descriptions. The Maidu thought of the Milky Way as Morning Star's path (Kroeber 1925:439), while the Pomo called it Bear's path (Loeb 1926:228-229). Generally, however, it was considered to be the path taken in a sky race, such as between Deer and Antelope, Coyote and Wildcat, or Falcon and Duck; this was true of the Monache (Gayton 1948:267), Yokuts (Gayton 1948:89, 156, 162, 230), Kawaiisu (Driver 1937:86-87), and the southern California Cahuilla (Hooper 1920:363). The Tipai called the Milky Way "Sky's Backbone" (Drucker 1937a:26).

Orion

Most, if not all, of native California recognized this constellation, particularly the belt stars and sometimes the sword and other elements. In northern California, the Yana considered the belt to be Coyote's arrow (Sapier and Spier 1943:283). The Klamath recognized the belt and the sword, but the identifications themselves are unknown (Spier 1930:221).

The belt stars were recognized by central Californians, including the Salinan, Chumash, Kitanemuk, Yokuts, Monache, Kawaiisu, and others; often they were identified as three men, but sometimes more (Driver 1937:86-87; Harrington 1942:29-30; Hudson and Underhay 1978:107-109). In the eastern and southern parts of the State, such as among the Paiute (Kelly 1932:154) and Tipai (Spier 1923:319-320, 358), a definite association of the belt with mountain sheep is present. The Ipai attributed some faint stars to one side as representing the animal's horns, while other stars within Orion were the hunters and their arrow (DuBois 1908:165). The Cahuilla, Cupeño, Luiseño, and Ipai stated that the stars represented a left-handed hunter with arrow (Drucker 1937a:26). The Chemehuevi identified the belt with mountain sheep, the sword with an arrow, and the three stars above the belt as the arrowhead. The two stars below, one of them Rigel, were the hunters (Laird 1976:91-93). A similar identification was made by the Kamia, who saw the belt stars as an antelope, a deer, and a mountain sheep, and the hunters as a Chemehuevi and a Yavapai (Gifford 1931:66-70). The Cahuilla identified one hunter, Rigel (Hooper 1920:363).

These stars were everywhere important, and in much of southern California they were mentioned together with the Pleiades (DuBois 1908:162-164; Harrington 1933:200-201; Hudson and Underhay 1978: 107-109). They were commonly represented in art.

The Pleiades

Throughout much of California these stars were identified as sisters, wives, maidens, or girls. The Atsugewi viewed them as seven sisters (Garth 1953:195), while the Yokuts and Monache named them as six young girls or women (Driver 1937:86-87; Gayton 1948:89, 156, 162, 230, 267; Latta 1936: 139-143). The Salinan, Chumash, Kitanemuk, and Gabrieliño termed them maidens (Harrington 1942:20-30; Hudson and Underhay 1978:109-112). In southern California, the Cahuilla, Cupeño, Luiseño, Ipai, and Tipai recognized them as seven in number (DuBois 1908:162-164; Spier 1923:319-320, 358; Gifford 1931:66; Drucker 1937a:26). A Cahuilla myth, however, reports them to have numbered only three (Hooper 1920:363).

Other identifications also existed. Among the Klamath they were little children (Spier 1930:221), while in northwestern California they were considered six persons in a boat (Driver 1939:344). The Costanoan perhaps also considered them as children, calling them the "bunch of little ones shaking" (Harrington 1942:29-30). The Chemehuevi and Kawaiisu identified them as Coyote's family, the six stars being his wife, a son, three daughters, and a shaman (Driver 1937:86-87; Zigmond 1977:72; Laird 1976:91-93). The Wintu, Nomlaki, and Nevada Shoshone also recognized these stars, but their mythological identification is unknown (DuBois 1935: 75; Steward 1941:267; Goldschmidt 1951:389).

These stars were represented frequently in southern California art.

Polaris

The Wintu (DuBois 1935:75), Nevada Shoshone (Steward 1941:267), Yokuts, Monache, and Paiute (Driver 1937:86-87) identified this star. Among the Pomo it was considered the eye of the creator, *Marumda*, who watched over the people below (Loeb 1926:228-229). The Chumash recognized Polaris as one of the creators, Sky Coyote (not to be confused with Coyote), and addressed to it prayers concerned with the welfare of people (Hudson and Underhay 1978:100-102). One informant stated that men would rise about midnight and recognize that it was the right time to go outdoors and talk to certain stars and ask for long life (Harrington n.d.).

In southern California, the Yuma considered the North Star as one end of Coyote's net. The stars which revolved around Polaris (Ursa Major, Ursa Minor?) were seen by the Cahuilla, Cupeño, and Ipai as Wildcat's rump (Drucker 1937a:26).

The Luiseño believed Polaris to be a great star chief whose people moved around him; that is why they danced in a certain direction in ceremonies, to imitate the movement of these stars (DuBois 1908:162-164). On a very clear night the star chief's heart and three fingers of his hand could be traced among several tiny sixth magnitude stars. One finger was missing because it had been bitten off by a rattlesnake. The tip of the heart was said to reach the horizon. This group of stars was called "Heart of the Wolves"; therefore Polaris was probably identified as Wolf. The Ipai represented Wolf as a star in their ground paintings in a northern position (Waterman 1910:352-353). Kroeber (1925:682), however, states that this identification is wrong, and that the figure in the ground painting is (Sky?) Coyote and not Wolf. If Kroeber be correct, an obvious parallel exists between the Luiseño Coyote Clan ceremony which focused upon Polaris in prayers, orientation of ceremonial participants to this star, and imitations of it by medicine men and dancers alike (Dougan 1964:142, 144).

Of all the stars, none exceeded the ritual importance of Polaris among central and southern Californians.

Scorpius

The Ipai recognized this constellation as a

boy with bow and arrow (DuBois 1908:165). This may be the group of stars which Gifford suggests equated with the Kamia and Tipai identification as *Chiyuk* or *Ciluk* (see Leo).

Spica

This star was considered a star chief among the Luiseño (DuBois 1908:162-164), but its mythological identity is unknown.

Ursa Major

Native Californians everywhere recognized these stars, but there is variation in mythological identifications and the number of stars mentioned. The desert groups in eastern and central California --the Paiute, Nevada Shoshone, Kawaiisu, and others --viewed these stars as either a rabbit net or people driving rabbits into a net (Kelly 1932:154; Steward 1941:267; Driver 1937:86-87). In central California, the Chumash, Monache, and probably the Yokuts, saw the constellation as seven boys (Gayton 1948:267; Driver 1937:86-87; Hudson and Underhay 1978:105-106). Chumash mythology connects the constellation with the story of seven boys turned to geese; these boys were honored in a ritual which took place in the fall when Ursa Major could be seen "standing on the handle," a configuration which may be analogous to the flying formation of geese. The Gabrieliño probably also recognized these stars as geese, since they associated the north with these birds. In one of their myths a man was tested by the elite by requesting him to draw a map (ground painting?) showing the exact relative position of these stars to Polaris (Heizer 1968:65-66, 129-130).

In northern California, the Nisenan considered the bowl stars to be three girls and a boy (Beals 1933:357). According to the Klamath, these stars were loons who gambled by diving into a lake to see who could dive the deepest (Spier 1930:221). The Pomo and Wappo saw them as a long pole with a hook (Loeb 1926:228-229; Gifford and Kroeber 1937:190; Driver 1936:195). The Atsugewi called them Coyote's cane (Garth 1953:195), which suggests that perhaps these people identified Coyote with Polaris. The Maidu knew these stars as "looking around" (Kroeber 1925:439), which is analogous with Pomo "eye of the creator."

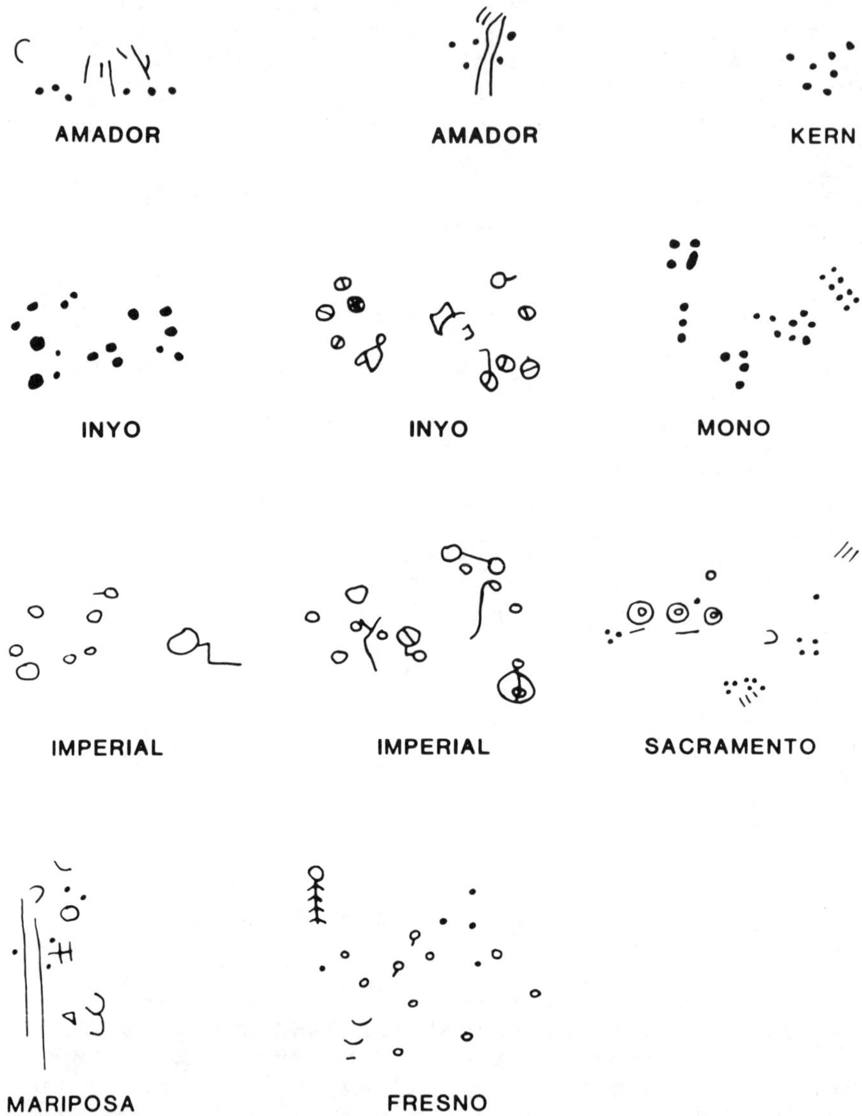

These other rock art patterns suggest depictions of stars, although they are not clearly identified with specific constellations. The crescent forms in these depictions might, therefore, be lunar symbols. (after Heizer and Clewlow, 1973, II--figures 1h, 5c, 9a, 19a, 19b, 32i, 39j, 78g, 129i, 171j, 208j. Griffith Observatory)

Ursa Minor

Little is known about this constellation. Hudson and Underhay (1978:104) report that it was recognized by the Chumash and may have been associated with Fox.

Vega

The Luiseño identified this star as the raised right hand of the Buzzard chief, Altair. They believed that its emergence before Altair signaled the chief's coming (DuBois 1908:162-164). The Chumash probably identified Vega as the stinger on the tail of Scorpion Woman, a constellation through which the dead must travel to reach Altair in the Land-of-the-Dead (Hudson and Underhay 1978:120-121)

Unidentified Stars

As might be expected, there are many stars and constellations described by native Californians for which identifications with our patterns are lacking.

One constellation which appears to have been widely recognized was named "Husbands of the Pleiades." It is mentioned by the Yokuts and Monache (Gayton 1948:89, 156, 162, 230, 267). The Cahuilla also named certain stars and constellations near the Pleiades as being mythologically connected with them. The brothers *Isilihnup* and *Holinach* are examples, each described as carrying a burning stick. *Tukwishhemish* is another star mentioned as being near the Pleiades; it is a large woman with her arms outstretched. The Pleiades follow her and try to look at the woman, but she keeps turning away. *Kunvachmal* is a bright star that rose only a little above the horizon and was identified as a half-brother of *Isilihnup* and *Holinach*. Another star, *Tukvachtahat*, was their father. *Papinut* was the "funny star." It was placed by *Wiyot* to be funny. It rose only a little above the horizon and twinkled more than the others (Hooper 1920:363-366).

The Tipai recognized a "Jealous Star" (Spier 1923:319-320, 358). The Chemehuevi identified four dim stars as "Jackrabbit's Setting Place," where he had been pursued by Wildcat (another star?); a set of three or four dim pairs of stars were attributed as "Jackrabbit's Tracks." The Chemehuevi also watched a spring constellation they called *Isagawi*, and another called "Broken Wing Duck" (Laird 1976:91-93).

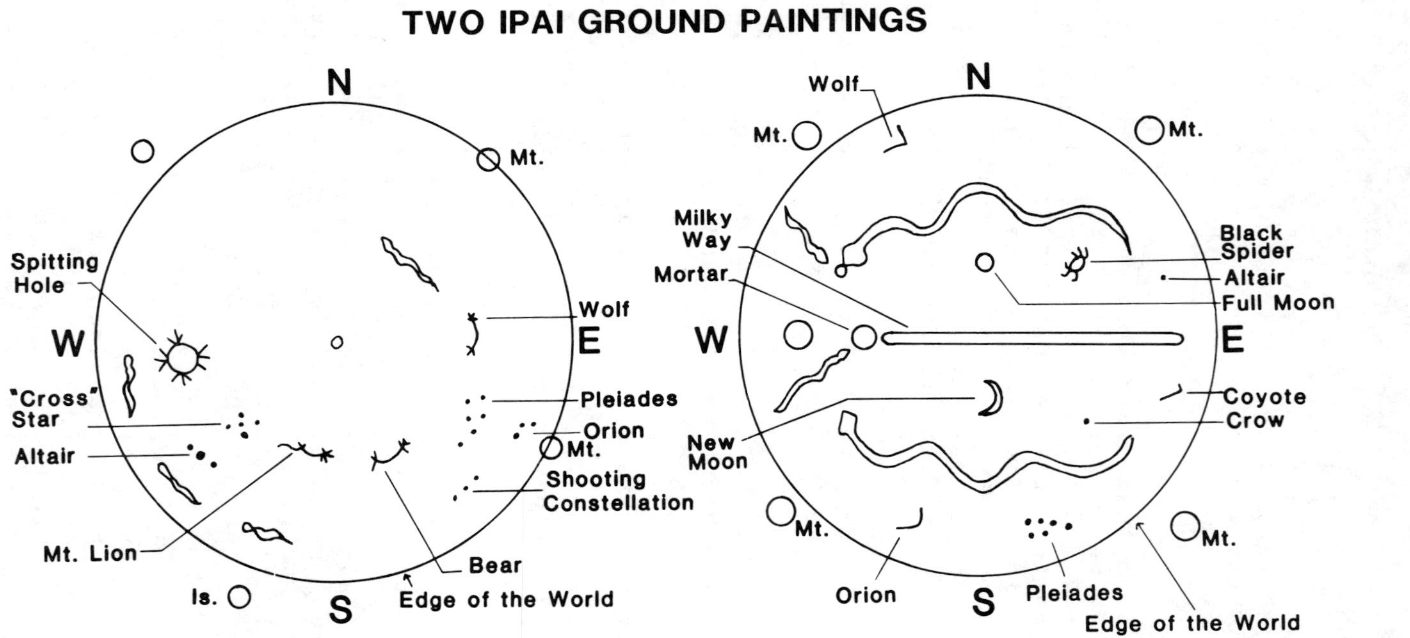

Two Ipai Indian ground paintings incorporate numerous celestial symbols into their designs, which seem to represent the cosmos. (after Waterman, 1910. Griffith Observatory)

Among the Sky People recognized by the Paiute were Eagle, Weasel, Flies, Two Women Fighting, and Big Woman (=Cahuilla's *Tukwishhemish*?) (Kelly 1932: 154). The Monache and Yokuts reported Blackbird, Raven, Roadrunner, Dog, Duck, Hummingbird, and Y-Frame Cradle (Driver 1937:86-87, Gayton 1948:89, 156, 162, 230). The Chumash mentioned Lizard's Hand, Raccoon (Alcor?), and Coyote's Son, to name a few (Hudson and Underhay 1978:103, 106). The Pomo identified a summer star as "Hungry Star"' it was often seen in front of Morning Star (Gifford and Kroeber 1937:190).

The juxtaposition of certain stars near Moon was also observed, and throughout native California the event represented an ill omen. In the northeast, the emergence of a single star close to Moon was considered a sign that sickness or death would occur (Voegelin 1942:146-147). The Chumash held a similar belief, noting that when a small star appeared three or four days after the new moon and near its lower point, an approaching disaster was indicated (Hudson and Underhay 1978:76).

Stars were depicted by the Chumash on various cult objects (ceremonial whistles and tubes, dancing skirts, and solstice sunsticks), and in ceremonial bodypainting and rock art. Star maps, consisting of stone tablets smeared with black tar in which white beads had been pressed, have been described by an informant as being used by certain men to determine the seasons. A complete representation of the sky from Orion's Belt to Altair has been found on one Chumash ceremonial whistle (Hudson and Underhay 1978:130-133). Recently I have seen three additional Chumash whistles in a private collection, one identical with the one just mentioned. The other two have depictions of the Pleiades. Star depictions in Ipai ground painting have been known for some time, but their counterparts in rock art have not been found. Mayer (1977), however, has postulated depictions of sidereal lunations, solstice markers, and perhaps even the Crab supernova of A.D. 1054, in the rock art of east-central California.

Calendrics

Although methods of time-reckoning varied from group to group, calendars served to integrate people with their natural and supernatural surroundings. The calendar regulated the timing for important events, such as activities involving economic, political, or ritual affairs. Sensing that terrestrial and celestial changes were connected, and imagining these changes to be caused by the actions of various beings and of people with power, supported and gave a sanctification to the need for reckoning time. In a meaningful way it provided understanding of the total environment and a way to predict change. A Cahuilla chief expressed it this way:

> When the sun swung to the north and the moon showed quartered by day overhead, or west, they knew by the signs of the sun and the moon when the seeds of certain plants were ripe, and they got ready to go away and gather the harvest. Every plant that grew, the nesting time of all the birds, the time for the young of all the animals, the time of the young eagles, everything they learned by the signs of the sun and the moon.
>
> They studied the north star, how it turns about, and the seven stars, and the morning star--all this helped to know when to go and gather their food. The month that the road runner flies means certain things, and the habits of many animals all meant something to these older people who studied the signs of the sun and the moon and the stars and the animals [Patencio 1943:113].

The basic unit of time was the day and its subdivisions were determined by the sun during the day and the stars at night. The Klamath, for example, used Orion's belt to tell time in the winter sky. Bedtime came for listeners of tales told on winter nights when these stars were two-thirds of the distance to the zenith (Spier 1930:218). The Pomo watched the changing position of Ursa Major to schedule their nightly fishing activities. They could tell what time a certain kind of fish was most likely to bite (Loeb 1926:228-229). Nightly rituals were also timed by the stars, a fact reported for the Chumash (Hudson and Underhay 1978: 119) and Cahuilla (Strong 1929:129).

Monthly time reckoning was based upon a lunation. Each month was given either a numerical designation (such as tenth moon or index finger moon), or a descriptive name (such as month when seeds ripen, or month when leaves fall). Broader descriptive names corresponding to the seasons served to group months together, and throughout most of California four seasons were recognized. As might be expected, there was considerable difficulty in equating a particular lunation with an annual natural event, because the lunar cycle does not correspond with the solar one. But by and large the sequence was corrected by observing the solstices, and an extra month was added as necessary. In many areas of California, especially in the north-central part of the state, arguments over the lunar months were typical (Cope 1919:140-143). Some designs in California rock art may be interpreted as tally marks for units of time; one such "calendar" has been reported by Schiffman (1977) for a Tubatulabal site. Another single panel seems likely for this region. Its right side appears to relate to the moon, the left to the sun.

A more subtle method of determining the year relied on the solstices. Clearly these observations were conducted by tribelets throughout the state (Hudson, Lee, and Hedges 1979:Table 1). Star markers were also widely used, but perhaps more so in central and southern California than elsewhere. This method was a way to monitor seasonal change, and it was more direct, since it was based on the earth's position as it orbits the sun.

In northwestern California, Wiyot astronomers kept track of the seasons by observing stars, such as the Pleiades (Driver 1939:342). The Yurok also recognized the Pleiades as a marker for the seasons; the cluster became invisible in its sixth month and was known to reappear in the dawn sky in the seventh (Kroeber 1925:75). Among the Shasta, these stars marked the beginning of summer (Holt 1946:341-342). The Pomo used the Pleiades and probably Orion's belt as star markers (Loeb 1926:229; Gifford and Kroeber 1937:148). The Pomo secret society knew that Ursa Major stood erect in the east in some months and upside down in others (Loeb 1926:228-229).

Star markers, especially the Pleiades, were also used in northeastern California. The appearance of a new moon between two stars at sunrise

POSSIBLE CALENDAR COUNT KER-114 KERN COUNTY, CALIFORNIA

What look like tally marks may make a calendar, or day count, out of this rock art panel from Tubatulabal territory. Based upon ethnographic data on calendrics presented in this study, the rayed disks on the left suggest to the author an association of this half of the panel with the sun. Similarly, the right half seems linked with the moon. The chain of circles on the far left may represent 8 years, and the 21 marks on the bottom may stand for a day count between new moon and the winter solstice in some year. The chain of circles that splits the panel could represent lunations observed in a particular year. The "tallies" on the right half number 16, 16, 28, 32, and 33, possible approximations to lunations or day counts related to lunations. The curved lines in one of these sets may refer to the crescent moon, with the bottom mark in the column meaning "new moon." (interpretation based on a tentative sketch, Heizer and Clewlow, 1973, II--figure 8. Griffith Observatory)

marked their fourth month and the time of winter solstice (Voegelin 1942:143). Spier (1930:217-221) stated that the Klamath did not use star markers. He added that they did mark winter when Gemini was barely seen over the horizon, and that other stars and constellations following Gemini were observed and named. Gemini's position overhead signaled the coming of spring. Several other stars on the horizon before dawn were also seasonal indicators; one seen in the east was identified with our month of May or June.

In central California, the Monache identified a particular star with late December, and they recognized the Pleiades as marking a summer month (Gayton 1948:267; Gifford 1932:31). The Yokuts used the Pleiades to mark the beginning of winter and as an indicator of how deep the rivers were to determine the spring salmon run (Gayton 1948:162). Star markers were also used by the Costanoan, Salinan, Kitanemuk, and Chumash (Harrington 1942:29; Hudson and Underhay 1978:133-135).

The reliance on stars to mark the seasons is better documented in southern California. The Kamia used the Pleiades, Orion's belt, and others (Gifford 1931:66-70). It was said that Wildcat had informed the people in the beginning of the world that they should plant their crops when the Pleiades appeared in the east, just before dawn. Among their non-agricultural neighbors, the Luiseño, a number of stars served as markers, and there were specific references to them in seasonal songs (Strong 1929:320; DuBois 1908:165; Harrington 1933:186-190). This practice was shared by the Serrano, Cupeño, and Ipai (Drucker 1937a:25-26; DuBois 1908:165). To determine the seasons, old people among the Cahuilla watched the stars, including the amount of light they cast. For example, in the second month of winter a particular set of stars was used. If they were dim, bad weather was forecast, but if bright, good weather lay ahead (Hooper 1920:365-366). Chemehuevi astronomers rose early to watch the predawn sky for star markers (Laird 1976:93-96). The Tipai noted the appearance of certain stars in relation to lunar phases to mark the months. The third month, for instance, was marked by Buzzard (Altair) rising when the moon was in last quarter; the fourth month was determined when *Ciluk* was seen in the predawn sky while the moon was new (Spier 1923:329, 357-358). Only the agricultural Mojave (and perhaps other farmers along the Colorado River) apparently paid little attention to star markers (Spier 1955:16-18).

The concept was widespread that the calendar had divine origins. In southern California the Tipai believed that Wildcat had instructed them in astronomy and calendrics, and that he had learned the information directly from the Sky People (Spier 1923:329, 357-358). The Kamia had a similar concept, that Wildcat was the repository of knowledge about the moons (Gifford 1931:65-66). Laird (1976: 157-158) published a charming Chemehuevi myth about the origins of the calendar. The First People gathered to discuss how many months there should be to a season. Coyote wanted each of the four seasons to have four months, but Burrowing Owl objected by simply holding up his little three-toed foot. Burrowing Owl later transformed himself into a rattlesnake, struck Coyote, and so won the argument. (Compare this account with the Luiseño description description of Polaris, in which it is said that one of Coyote's fingers is missing because he was bitten by a rattlesnake.) This Chemehuevi myth has a counterpart in northern California. The Tolowa said that there used to be fifteen moons which made the year too long. Coyote hid near their sweathouse and succeeded in killing two of them as they came out. The thirteenth moon managed to escape; Coyote only stabbed it in the rump (Drucker 1937b:240). Similar images from various regions of California suggest that the origins of the calendar and astronomical observations may have great antiquity. Like the New and Old World interpretations of the Pleiades as girls, "This may prove to be one of the concepts of independent or directly psychological origin which have so often been sought but are so difficult to establish positively" (Kroeber 1925:875).

The variability reported in native California calendar systems by Cope (1919:140-143) and Kroeber (1925:875) cannot be easily sorted into three types: descriptive, numerative, or astronomical. Native Californians clearly were aware of lunations and solstices, and they equated seasonal changes with the appearance of certain stars. All of these are attributes of astronomical time reckoning. Regional variations are probably differences of emphasis.

According to Cope (1919:140-143) the simplest type of calendar used in native California was the descriptive type, wherein natural seasonal events were equated with a sequence of lunations to establish the year. The Maidu year began in spring when flowers were in bloom. The pattern was similar among the Nisenan (Beals 1933:357-358), Nomlaki (Goldschmidt 1951:388-389), and Wintu (DuBois 1935: 68-70): There were four seasons and ten to thirteen moons which had descriptive names. For the Wintu the sequence began with the fall. A few old men made day and year counts either by planting notched sticks in a row or by placing pebbles in a basket. Reckoning years has also been reported for the Pomo and Chumash.

Although an obvious emphasis was placed on the moon by native Californians, we must remember that they also took an interest in the solstices. In actuality there were probably two calendars in use, a secular one based on the moon and known by the common people, and an esoteric one maintained by the elite and used for timing ritual activity.

Cope's (1919:140-143) second type of calendar was numerative. Its lunar periods are designated with numbers rather than with descriptive names. Sometimes this took the form of counting on the fingers of the hand. She reported this type used by the Modoc, but the Klamath can be added also. Four seasons were recognized by these people, and the year began in summer, or fall, as determined by various food resources.

The numerative calendar is basically a lunar calendar, but the Modoc and Klamath also recognized the solstices and star markers. Voegelin (1942: 143) notes that the Modoc and Klamath began the year with the winter solstice. This obvious contradiction suggests either that these groups had both a lunar and solar calendar (as suggested for the Maidu, Wintu, and others), or that both lunar and solar data had been incorporated into one calendar, but with the loss of solstice observers in historic times the solar component had been forgotten. The latter possibility may be true of the Maidu and their neighbors as well.

A numerative type of calendar based on the solstices was used in northwestern California. Cope (1919:140-143) reported that the Yurok used numbers to designate up to ten lunations with subsequent ones given descriptive names. The sequence began with the winter solstice. It has already been mentioned that the Yurok used star markers, such as the Pleiades and that their astronomers watched the daily change in the sun's position to determine the solstices. Driver (1939:342) reported that the head doctor or chief kept track of the months by tying a knot in a string to represent each lunation. These strings were hung from the sacred sweathouse roof. Both Cope and Kroeber considered the northwestern California calendar to be a blend of the numerative type and the third type: astronomical. This calendar is at least partly solar, and the distinctions between it and the Modoc-Klamath calendar do not seem to warrant treating either of them separately. It appears that before historic disruption, all of northern California employed a solar calendar with months designated by either numerative or descriptive names. The seasons were determined by the appearance of certain stars.

In central California, the Pomo used a solstice calendar with star markers; some tribelets named the lunar periods descriptively (Gifford and Kroeber 1937:148), and others numbered them (Loeb 1926:223-227). These people were also similar to their northern neighbors in that they used sticks to keep track of time. In fact, the Pomo were able to keep records up to 512 years! Loeb (1926:223-227) describes the tradition as follows. The head of the secret society watched the moons and used sticks to count them. A bundle of 13 sticks represented the year. There were three bundles of eight sticks each which represented 8, 64, and 512 years, respectively. These sticks were inherited and used to date important events, a facet of native California behavior which was previously unknown (Kroeber 1925:873). It is not specified what sort of events were recorded, but since the sticks relate to counting time and watching the sky, it may well be that the events recorded were things like lunar and solar eclipses, or perhaps the appearance or disappearance of Venus.

Giving months descriptive names appears to have a wide distribution in central California, including the Yokuts (Gayton 1948:89, 156, 162; Driver 1937:86), Monache (Gifford 1932:31; Gayton 1948:266-267), Nevada Shoshone (Steward 1941:267), Surprise Valley Paiute (Kelly 1932:152-154), and other groups. There is also an apparent interest in using both solstices to mark the calendar. While the use of a knotted cord to count the days is reported, two other techniques have also been described. The Surprise Valley Paiute marked the days as scratches on a stick, with an extra long one to designate the new moon at the beginning of the month (Kelly 1932:152-154). Apparently, the Tipai-Ipai also used this device, an example being currently on exhibit in the Museum of the American Indian. The Yokuts placed beads over a stick, feather, or bone, to serve as a calendrical mnemonic device (Driver 1937:86, Aginsky 1943:426).

The Costanoan, Salinan, Kitanemuk, and Gabrielino apparently also used a solar calendar, gave descriptive names to the months, and marked the year by the solstice as well as the stars (Harrington 1942:29). The Chumash calendar is particularly interesting because we know a little more about it. Each of its twelve descriptive months was probably marked by a star and had astrological meaning. An individual born in the Chumash month equivalent to our September was said to be "careful and watchful of dangers"; a person born in February would be a "victim of uncertainty and never sure of anything" (Blackburn 1975:Narrative 13). These personality attributes were apparently considered to be determined by birthdate, and perhaps they were derived from the celestial beings whose powers were in force at that particular time (Hudson and Underhay 1978:126-136). This intellectual association of people with stars has obvious parallels in Old World astrology. Perhaps the Chumash used it to decide upon those best qualified to assume important ritual and political offices.

Hudson and Underhay (1978:138-139) have considered the problem Chumash astronomers faced in keeping their twelve months of 30 days each correct with the solar year and in agreement with their seasonally descriptive names. Evidence on method is uncertain but indicates that once every five years an extra month was added. Ideally, the winter solstice and the new year should come at the same time as a new moon, but since this seldom happened, the number of days from new moon to winter solstice must have been counted. The belief in Moon as scorekeeper in the nightly gambling game, the outcome of which was determined at the time of winter solstice, may in fact have been a form of day count. It may be speculated that if the new moon came at or shortly before winter solstice, the idea of Moon's giving life at this time of month was extended for the entire year. But if the moon were waning, the prediction would be death. Like the Wintu and Pomo, the Chumash also counted years.

The interest in both solstices (as noted earlier for the Yokuts, Monache, and Chumash) was highly emphasized in the southern California calendar, which Cope (1919:140-143) considers her third type: astronomical. Among the Juaneño, Luiseño, Ipai, and Tipai, the calendar utilized both solstices, as well as star markers to determine the months (Kroeber 1925:873-875). The months were thus fixed, but only those for half the year were named. These names were apparently repeated for the second half of the year. Lunar and stellar observations were important in determining the months. It has already been mentioned that the appearance of a certain star was noted when the moon was in a certain phase (Spier 1923:329, 357-358; DuBois 1908:165; Strong 1929:320). The pattern seems to have been used by the Cahuilla (Hooper 1920:362) and Kamia (Gifford 1931:65-66), and probably also by the Serrano, Cupeño, and Chemehuevi (Drucker 1937a:25-26). Whether it was present among the Mohave is uncertain (Spier 1955:16-18). This southern California form of solstice calendar is clearly allied to that of the Southwestern Pueblo peoples, while the northwestern California form is connected with that of the Pacific Northwest Coast peoples (Kroeber 1925:874).

All of these calendrical systems for native California clearly emphasize the observance of the solstices and the use of star markers. Because of the cyclical nature of the solar year and the seasonal round, the economic importance of the calendar is apparent. But the connection between human beings and celestial beings involved much more than simple economics, for a calendar was certainly intended to include rituals which focused upon a person's successful passage in death from this world into the upper one. These rituals, or Mourning Ceremonies, took place every few years and required the powers of people to help the souls of those who had died in the interim to reach the Land-of-the-Dead. Because the upper world was filled with powerful beings (celestial objects) and considered an extremely dangerous place for people, it was essential to schedule such ceremonial activity when the circumstances were best for human success. Therefore, the ritual calendar, which served to order celestial events as well as earthly ones, was important in planning various religious activities by determining which celestial being's powers (positive, negative, or neutral) people were most likely to encounter.

At the time of the Mourning Ceremony, one being which was certainly encountered was Evening Star, and probably also Altair, the captain who ruled over the Land-of-the-Dead. The solar calendar (which used Altair as a star marker) would enable astronomers to schedule rituals which required Altair's presence in some particular location in the sky. But when a planet is also important to the ritual, a system to reckon lengths of time greater than a year is necessary. There would be a need for a ceremonial calendar whose maintenance included the recording of planetary observations (particularly Venus) as well. It is known that Wintu, Pomo, and Chumash astronomers maintained year counts, and they may well have done so to keep track of a few planets, especially Venus.

We do not know if central and southern Californians were keeping track of Venus, but it can be inferred that they were. Certainly they were keenly aware of the daily celestial changes and used these to explain earthly changes. Some planets, especially Venus, were also observed and worshipped, but the "behavior" of this object could not be equated to definite seasonal changes in nature. Rather, they may have perceived this planet as affecting earthly conditions in general, transcending the four seasons, introducing unpredictable qualities to the universe, and also affecting human conditions, especially life and death. The Yokuts Mourning Ceremony, described as occurring as Evening Star passed into inferior conjunction, may therefore have been deliberately scheduled to coincide with this celestial event. The means of timekeeping required to anticipate such events were certainly present, as were astute observers who believed the events they saw in the heavens held the very destiny of their people.

The question of whether native Californians were able to predict the movements of Venus cannot be answered by the ethnographic data now available. Laird (1976:94) provided the reason for the many gaps in our knowledge of native California astronomy when she wrote:

> When there were no longer youths eager to receive the ancient lore...the old men stopped their futile watching, and their knowledge perished with them.

The Astronomers

The individuals who undertook all of these observations of the sun, moon, planets, and stars to reckon time were the elite. In northwestern California the task belonged to the head doctor or chief (Driver 1939:342). Among the Pomo, it was the chief who kept records of the moon, determined the solstices, and recorded time (Driver 1936:195-196; Gifford and Kroeber 1937:148). Loeb (1926:229) stated that the esoteric variety of astronomical knowledge was given to members of the Pomo secret society, whose responsibility it was to address certain prayers to the sacred stars, and to time these prayers based upon their observations.

The astronomers in other northern groups were said to be simply old men by the Nisenan (Beals 1933:357-358), Wintu (DuBois 1935:68-70), and Nomlaki (Goldschmidt 1951:388-389). In the northwest the old men were identified as three or four doctors who watched the moon, daily changes of the sun at sunrise and sunset, and the stars, and who kept time records (Voegelin 1942:143).

In central California old men astronomers were mentioned by the Miwok, Yokuts, and Monache (Aginsky 1943:426). Among the Chumash these men belonged to a secret society, the *'Antap*, whose leader held the title of Sun Priest. Its twelve members, each perhaps representing a given Chumash month, engaged in observations and rituals in accordance with the stars. They used the term

'Alchuklash to designate their astronomer shamans (Hudson and Underhay 1978). Each may have been a specialist whose powers were necessary when his particular month's star appeared. A Spanish priest wrote in 1803 that all events, even natural ones, were believed to be caused by witchcraft and poisons (=power), "and that the scarcity of rain and of seeds proceed from certain men who among them are regarded as gods of the waters, and of the good or bad crops" (Engelhardt 1932:6-8).

Actions taken by one of these Chumash specialists, the rain doctor, will be of interest to researchers concerned with the initial stages of plant domestication:

> At new moon, in month 'when icicles dropped from house' (January), rain doctor filled little black stone bowl with water and planted whole acorn, chia seeds, every other kind of seeds in this bowl and left it inside his house for month; He did this to make all the seeds grow well that spring. He talked over these seeds, the same way that he talked when he went outside to make rain. [Voegelin 1938:64].

(A similar practice has been reported for Kawaiisu weather shamans; during rainmaking ceremonies, seeds were placed in a special stone bowl filled with moist mud and, although we are not told what became of them, perhaps were set aside to germinate (Zigmond 1977:88). For the Chumash, there is evidence for other special duties being performed to the calendar, perhaps undertaken monthly by each *'Antap* as his particular star emerged (Hudson and Underhay 1978:135-138).

It is probable that a similar concept existed throughout much of southern California. Fray Boscana (1933:29) observed that among the Juaneño and Luiseño certain individuals had the powers to make rain, influence the dew, or produce acorns. These individuals were also members of a secret society whose origins and affinities were with the Chumash *'Antap* cult (Hudson and Blackburn 1978). Elite specialists were involved in astronomical matters among other groups, such as the Kamia (Gifford 1931:65-66), Tipai (Spier 1923:329, 357-358), and Chemehuevi (Laird 1976:93-96). The astronomers among this latter group were described as coming out of their houses daily to stand naked in the predawn cold to make their stellar observations. Like the Chumash, the Cahuilla had a council of elders who were the astronomers:

> The old men used to study the stars very carefully and in this way could tell when each season began. They would meet in the ceremonial house and argue about the time certain stars would appear, and would often gamble about it. This was a very important matter, for upon the appearance of certain stars depended the season of the crops. After several nights of careful watching, when a certain star finally appeared, the old men would rush out, cry and shout, and often dance. In the spring, this gaiety was especially pronounced, for it meant that they could now find certain plants in the mountains. They never went to the mountains until they saw a certain star, for they knew they would not find food there previously [Hooper 1920:362].

In view of the apparent connection between native California clans and their celestial counterparts among the Sky People (Hudson and Blackburn 1978), it can be speculated that these astronomer specialists may have been the heads of clans and moieties. In support of this are native statements about the origins of clans and moieties; for example, the Cahuilla believed that Moon assigned the various people to their clans (Strong 1929:136). Conceivably, it required the total unity of these various social groups to maintain the balance to the universe. As celestial changes took place in the heavens above, each astronomer's powers waxed or waned and led to arguments among themselves, but everywhere these specialists were in complete agreement that their activities were necessary ones to maintaining a balanced and viable cosmos.

Summary and Implications

California's first astronomers compiled a body of astronomical information and devised explanations for what they observed in the sky to an extent far beyond what has been previously credited to hunter-gatherers. Their involvement with the heavens rested upon a complex philosophical framework of power, which included celestial objects, people, and other beings. Because celestial objects were perceived as supernaturals, capable of acquiring and using power to alter the universe, it was essential for people to watch them in earnest, and not for mere entertainment. Throughout California, astronomers observed the motions, positions, "behavior", and characteristics of these beings in an effort to understand and to cope with the changes they signaled in the natural and psychological environments. Astronomers were not helpless against the forces of the universe. Ever since people were created, it had been their responsibility to acquire power of their own and to use it to maintain the balance of the cosmos. But the universe was dynamic, and to cope with its ever-changing conditions the astronomers were also given an indicator to anticipate cyclic changes in the use of power by the celestial beings. This indicator was the calendar, which established a bond between man and gods and applied a system of time-ordered ritual activity and understanding to the actions of sun, moon, and stars.

Those celestial beings whose power was greatest received the most attention in observation and ritual. Sun was one of these, since its movements across the heavens were linked directly to seasonal changes. Everywhere its powers were considered enormous. Human responsibility included delivery of daily prayers. Sun had to be watched closely and, when the solstices drew near, the rituals that properly ensured the all-essential cosmic balance were undertaken. Daily ritual activity probably was performed by the astronomers and perhaps at observatories near major communities. Throughout much of central and southern California the biannual and annual rituals to the sun were important community events. They involved the entire tribelet.

Moon was also of great importance. It was believed to influence the health of the community and to exercise great influence upon various natural rhythms, such as menses and tides. The moon's phases were closely watched. The new moon was a time of ritual activity on the part of the community. Daily counts of the moon were maintained by the astronomers.

Stars, particularly very bright and conspicuous ones, were also important and, like Sun, caused seasonal changes on earth. Some served as markers to indicate these changes, and for central and southern Californians they were extremely important calendrical tools. Of all the stars equated with celestial beings, Polaris was the most important. Its central position in the heavens and constant presence gave a strong sense of stability to the universe; lesser beings circled Polaris, emphasizing its importance. This star was identified in many areas of California as a benevolent being responsible for a creative role in the universe; it may have been widely identified with Sky Coyote. He was also linked to the creation of the calendar itself by determining how many moons should make up the solar year.

Throughout all of native California the astronomers watched the stars and discussed them as signals of seasonal change. Unfortunately, the body of knowledge about various stars and native ordering of them into constellations is poorly understood. It can be said, however, that many stars were recognized, particularly in southern California, where even very faint sixth magnitude stars were identified with specific constellations, as for example Wolf's, or Coyote's, Heart among the Luiseño.

All of these observations were coupled with the calendar, and they enabled detection of power shifts within the year. Therefore, the necessary ritual activity could be performed at the proper time to ensure a stable, economic resource base. In spring, observations were made and prayers were addressed to certain stars with the belief that their appearance meant that power would be exercised to stimulate the growth of various plants and animals. The frequent statements by native Californians that their astronomers would engage in arguments over whether or not a given star had appeared suggests that it was the astronomers themselves who were receiving and using power provided by these celestial objects. Conceivably then, individual astronomers identified themselves with specific celestial beings whose powers they used. As these objects appeared or disappeared, so did an individual's power. There is also a probable connection between astronomer and clan leader. The clan's identity may have been mythically derived from a given celestial being.

The struggle for power between astronomers was probably rooted in the same struggle between the celestial beings themselves, as each attempted to gain superiority over the others. Most celestial struggles were predictable, for example, the changing phases of the moon and the approach of sun toward solstice. Other struggles were not predictable but could be explained, and they too required ritual interaction and the use of power by people. Among the most terrifying unpredictable events were solar and lunar eclipses and the appearance of a comet. A hint that some preparation for ritual activity took place for eclipses suggests the ability to predict these celestial struggles,at least some of them.

It is uncertain whether Venus's movements were predicted. Throughout native California this planet was observed daily, and prayers to it to influence Sun or Moon indicate its great power. In central and southern California, it appears that the planet, seen as two separate objects which alternated in appearance and disappearance, was identified as two separate beings. Venus as Morning Star was conceived as benevolent, but Venus as Evening Star was malevolent. The latter, at least among the Chumash, and perhaps the Gabrieliño, was equated with death.

To a people making constant daily observations of the sky and explaining changes taking place around them as a result of the exercise of power by celestial beings, it was logical to relate these shifts in power to the solar year. Therefore, the appearance of a given star or the sun at a particular point on the horizon signaled the predictable changes about to take place on earth. But Venus--a bright and conspicuous object considered to have powers greater than stars--as either Morning or Evening star provided no definite, predictable influence over seasonal events. As a consequence the astronomers considered its use of power either to be more erratic and emotional than the other beings, or (at least in parts of central and southern California) to be focused upon influencing people rather than nature. The latter view is suggested by association of the object with life and death, and in particular by the prospect that the Mourning Ceremony for the dead may have been held when the planet disappeared as Evening Star and later reappeared as Morning Star.

It is known that the upper world was believed to be dangerous and powerful, and that people timed their interactions with its beings to coincide when they had the greatest potential for success. In the Mourning Ceremony, this action on the part of a person was necessary to aid the soul in successfully reaching the Land-of-the-Dead (Lowell Bean, personal communication). It therefore follows that the Mourning Ceremony was probably timed to coincide with the presence of celestial beings who would be helpful toward human efforts to aid the soul on its journey. In this case the passage of Venus into inferior conjunction could very well have been extremely important.

But if ceremonial activity involving the dead required such critical timing, then some native Californians--particularly those in central and southern California where the Mourning Ceremony was practiced--must have developed a ritual calendar which predicted when Venus would be making the transition from Evening to Morning stars. It is suggested that a ritual calendar, from Mourning Ceremony to Mourning Ceremony, was in use and incorporated into the round of ceremonial activities connected with the solar calendar. The astronomers observed Venus daily, and by the use of stick bundles or other devices were able to maintain counts of days for periods longer than the solar year. In the case of the Pomo, such records reached up to 512 years and were inherited, most likely, by one astronomer from another.

While we do not yet know all the mechanics of native California astronomy, the justifications are readily apparent. To understand and to cope with change, it was necessary to understand and cope with the forces responsible for it: the celestial beings themselves. For an outdoor people highly observant of their surroundings, these connections were logical. That other hunter-gatherers in the more ancient world may have reached similar conclusions is entirely possible, particularly those people whose interaction with the natural environment resulted in domestication of plants and animals and the formalization of knowledge into a "science," religiously based.

The California data clearly indicate that the roots of science, such as the origins of calendrics and astronomy, are not invariably the intellectual achievements of food producers. Our awareness of the sky and our relation to it are important philosophical concerns in interacting with the total environment. What can be seen for native California holds important consequences for our understanding of the origins and developments of other astronomies and calendars among food producers:

• Knowledge of the motions of celestial objects and how they equate with seasonal change is one such step toward the development of food production The first experiments with wild plants and animals were undertaken by hunter-gatherers who may well have already achieved a level of understanding in regulating economic activities by the appearance of various celestial objects, and who arranged these activities in the form of a solar calendar. The major difference between a hunter-gatherer calendar and an agricultural one is--perhaps--the need to determine the equinox, thereby providing greater refinement to a time-reckoning tool already capable of allowing man to predict when certain climatic changes will occur.

• An attempt to understand the movements of the planets, which were believed to be influential over people and nature but otherwise unpredictable within the solar year, led to development of knowledge about these bodies as a means to gain even more control over the environment.

• The use of day, month, and year counts to record astronomical events would lead to the ability to understand and predict planetary cycles and eclipses, and to place these in terms of solar dates. This step would logically follow into the development of mathematics, and in the case of the Pomo use of various stick bundles to represent numerical relationships, a form of positional notation.

• The group of elite whose purpose it was to observe, record, interpret, and conduct ritual activities based upon celestial events unquestionably would develop into full-time astronomer specialists who would also form the priesthood by which the society's interaction with the gods was assured.

All these aspects of our attempt to understand and to control our environments were highly developed among the ancient civilizations of Mesopotamia Egypt, China, Mesoamerica, and Peru, and were no doubt further refinements of a body of knowledge used by their food producing antecedents as well. That these aspects may have had their roots even earlier among their hunter-gatherer ancestors now seems probable too.

Acknowledgments

I would like to express my appreciation to Drs. Dennis Power, Lowell Bean, Maurice Zigmond, and Thomas Blackburn, and to Mrs. Janice Timbrook, for their many helpful comments and suggestions on this paper. Also, my thanks are extended to Mr. Clifton Smith and Mrs. Joan Pursell of our Museum library for locating books and articles. Most of all, I would like to express my sincere appreciation to Dr. Edwin Krupp for inviting me to participate and to the American Association for the Advancement of Science for its support in making this paper a reality.

Notes

[1]When the terms *sun* and *moon* are capitalized in this paper, I am referring to these objects as supernatural beings. The same is also true for various beings identified with other celestial objects (for example, "Morning Star").

References

Aginsky, B.W.
1943 Culture Element Distributions: XXIV: Central Sierra. University of California Anthropological Records 8(4):393-468.

Barnett, H.G.
1937 Culture Element Distributions: VII: Oregon Coast. University of California Anthropological Records 1(3):155-204.

Barrett, S.A.
1919 Myths of the Southern Sierra Miwok. University of California Publications in American Archaeology and Ethnology 16(1):1-28.

Beals, Ralph
1933 Ethnology of the Nisenan. University of California Publications in American Archaeology and Ethnology 31(6):335-414.

Bean, Lowell
1974 Social Organization in Native California. In 'Antap: California Indian Political and Economic Organization, Lowell Bean and Thomas King (editors), pp. 11-34. Ramona: Ballena Press Anthropological Papers No. 2.

1975 Power and Its Applications in Native California. The Journal of California Anthropology 2(1):25-33.

Blackburn, Thomas
1963 A Manuscript Account of the Ventureño Chumash. University of California, Los Angeles, Archaeological Survey Annual Reports, 1962-1963, pp. 135-158. Los Angeles.

1975 December's Child: A Book of Chumash Oral Narratives. Berkeley: University of California Press.

Bock, Frank, and A.J. Bock
1974 Western petroglyphs. KaiDib Films International, Glendale.

Boscana, Fray Geronimo
1933 *Chinigchinich*. Santa Ana: Fine Arts Press.

Cook, Sherburne F.
1978 Historical Demography. In *Handbook of North American Indians*, Vol. 8: *California*, Robert Heizer (editor), pp. 91-98. Washington, D.C.: Smithsonian Institution.

Cope, Leona
1919 Calendars of the Indians North of Mexico. *University of California Publications in American Archaeology and Ethnology* 16(4):119-176.

Davis, Edward
1919 The Diegueño Ceremony of the Death Images. *Contributions from the Museum of the American Indian, Heye Foundation*, Vol. V, No. 2. New York.

Dougan, Marjorie
1964 "The Memorial Ceremony of the Luiseño Indians." *The Masterkey* 38(4):140-149.

Driver, Harold
1936 Wappo Ethnography. *University of California Publications in American Archaeology and Ethnology* 36(3):179-220.

1937 Culture Element Distributions: VI: Southern Sierra Nevada. *University of California Anthropological Records* 1(2):53-154.

1939 Culture Element Distributions: X: Northwest California. *University of California Anthropological Records* 1(6):297-433.

Drucker, Philip
1937a Culture Element Distributions: V: Southern California. *University of California Anthropological Records* 1(1):1-52.

1937b The Tolowa and their Southwest Oregon Kin. *University of California Publica-*

tions in American Archaeology and Ethnology 36(4):221-300.

DuBois, Cora
1908 The Religion of the Luiseño Indians of Southern California. University of California Publications in American Archaeology and Ethnology 8(3):69-186.

1935 Wintu Ethnography. University of California Publications in American Archaeology and Ethnology 36(1):1-148.

Engelhardt, Zephyrin
1932 Mission Santa Ines. Santa Barbara: Mission Santa Barbara.

Garth, Thomas
1953 Atsugewi Ethnography. University of California Anthropological Records 14(2):129-212.

Gayton, Anna
1948 Yokuts and Western Mono Ethnography. University of California Anthropological Records 10.

Gifford, E.W.
1931 The Kamia of the Imperial Valley. Bulletin of the Bureau of American Ethnology, No. 97. Washington, D.C.: Smithsonian Institution.

1932 The Northfolk Mono. University of California Publications in American Archaeology and Ethnology 31(2):15-65.

1955 Central Miwok Ceremonies. University of California Anthropological Records 14.

Gifford, E.W., and A.L. Kroeber
1937 Culture Element Distributions: IV: Pomo. University of California Publications in American Archaeology and Ethnology 37(4):117-254.

Goldschmidt, Walter
1951 Nomlaki Ethnography. University of California Publications in American Archaeology and Ethnology 42(4):303-443.

Grant, Campbell
1965 *The Rock Paintings of the Chumash.* Berkeley: University of California Press.

Harrington, John P.
n.d. *Ethnographic Fieldnotes on the Chumash and on the Gabrielino.* Manuscripts, United States National Anthropological Archives. Washington, D.C.: Smithsonian Institution

1933 Annotations. In *Chinigchinich*, P.T. Hanna (editor), pp. 91-228. Santa Ana: Fine Arts Press.

1942 Culture Element Distributions: XIX: Central California Coast. *University of California Anthropological Records* 7(1):1-46.

Heizer, Robert
1968 The Indians of Los Angeles County: Hugo Reid's Letters of 1852. Los Angeles: *Southwest Museum Papers* No. 21.

Heizer, Robert, and C.W. Clewlow, Jr.
1973 *Prehistoric Rock Art of California. 2 vols.* Ramona: Ballena Press.

Holt, Catharine
1946 Shasta Ethnography. *University of California Anthropological Records* 3(4):299-349

Hooper, Lucile
1920 The Cahuilla Indians. *University of California Publications in American Archaeology and Ethnology* 16:315-380.

Hudson, Travis
n.d. *Archaeological Fieldnotes on SBa-1380, Indian Creek Area, Santa Barbara County.* Manuscript, Department of Anthropology, Santa Barbara Museum of Natural History.

1979 *Breath of the Sun: Life in Early California as Told by a Chumash Indian, Fernando Librado, to John P. Harrington.* Banning: Malki Press

Hudson, Travis, and Thomas Blackburn
1978 The Integration of Myth and Ritual in

South-Central California: The "Northern Complex." The Journal of California Anthropology 5(2):225-250.

Hudson, Travis, and Ernest Underhay
1978 Crystals in the Sky: An Intellectual Odyssey Involving Chumash Astronomy, Cosmology, and Rock Art. Socorro: Ballena Press Anthropological Papers No. 10.

Hudson, Travis, Georgia Lee, and Ken Hedges
1979 Solstice Observers and Observatories in Native California. The Journal of California and Great Basin Anthropology 1(1)39-63.

Kelly, Isabel
1932 Ethnography of the Surprise Valley Paiute. University of California Publications in American Archaeology and Ethnology 31(3): 67-210.

Kroeber, A.L.
1925 Handbook of the Indians of California. Bulletin of the Bureau of American Ethnology No. 78. Washington, D.C.: Smithsonian Institution.

Laird, Carobeth
1976 The Chemehuevis. Banning: Malki Museum Press.

Latta, Frank
1936 California Indian Folklore. Shafter: Shafter Press.

1949 Handbook of the Yokuts Indians. Oildale: Bear State Books.

Loeb, Edwin
1926 Pomo Folkways. University of California Publications in American Archaeology and Ethnology 19(2):149-404.

Marshack, Alexander
1972 The Roots of Civilization: The Cognitive Beginnings of Man's First Art, Symbol and Notation. New York: McGraw-Hill.

Mayer, Dorothy
1977 An Examination of Miller's Hypothesis. In Native American Astronomy, Anthony Aveni (editor), pp. 179-201. Austin: University of Texas Press.

Patencio, Francisco
1943 Stories and Legends of the Palm Springs Indians. Palm Springs: Palm Springs Desert Museum.

Sapier, Edward, and Leslie Spier
1943 Notes on the Culture of the Yana. University of California Anthropological Records 3(3).

Schiffman, Robert
1977 Rock Art of the Tubatulabal: Perspectives on Astronomy, Calendrics and Menstruation. Paper presented at the Annual Meetings of The Society for California Archaeology, San Diego.

Shipley, William F.
1978 Native Languages of California. In Handbook of North American Indians, Vol. 8: California, Robert Heizer (editor), pp. 80-90. Washington, D.C.: Smithsonian Institution.

Smith, Gerald A., and Wilson G. Turner
1975 Rock Art of Southern California. Redlands; San Bernardino County Museum.

Spier, Leslie
1923 Southern Diegueño Customs. University of California Publications in American Archaeology and Ethnology 20(16):297-358.

1930 Klamath Ethnography. University of California Publications in American Archaeology and Ethnology 30.

1955 Mohave Culture Items. Museum of Northern Arizona Bulletin 28. Flagstaff.

Steward, Julian
1929 Petroglyphs of California and Adjoining States. University of California Publica-

tions in American Archaeology and Ethnology 24(2).

1941 Culture Element Distributions: XIII: Nevada Shoshone. University of California Anthropological Records 4(2):209-259.

Strong, William
1929 Aboriginal Society in Southern California. University of California Publications in American Archaeology and Ethnology 26.

Voegelin, Erminie
1938 Tubatulabal Ethnography. University of California Anthropological Records 2(1): 1-123.

1942 Culture Element Distributions: XX: Northeast California. University of California Anthropological Records 7(2):47-251.

Waterman, T.T.
1910 The Religious practices of the Diegueño Indians. University of California Publications in American Archaeology and Ethnology 8(6):271-358.

Zigmond, Maurice
1977 The Supernatural World of the Kawaiisu. In Flowers of the Wind, Thomas Blackburn (editor), pp. 59-95. Socorro: Ballena Press Anthropological Papers No. 8

Archibald S. Thom

3. Solar and Lunar Observatories of the Megalithic Astronomers

Abstract

With passage of time, each generation has shown interest in the mysterious rows and rings of standing stones to be found in northwest Europe. Why were they erected, sometimes at great expense of the resources available to prehistoric people? We believe that at one stage in prehistory many of them were used in conjunction with the horizon to observe the rising and setting of the sun and moon. No written evidence exists; the evidence is only that obtainable from the stones themselves. Erected four thousand years ago, many of the stones seem to act as focal points in the landscape by involving the horizon, the sky, and the land. Each lichen-covered stone seems timeless, its own character formed by the weathering effects of wind, rain, and sun. We attempt to show that it is highly probable the erectors made and used a solar calendar and observed and recorded the moon's movements carefully, so carefully that they could have predicted eclipses. A sensitive method of observing the moon, one that even shows the effect of the sun's small perturbation (Δ) of the moon's orbit, could have been used by the people who set out the stones.

Astronomical Concepts

The megalith builders could make and record *accurate* observations of the azimuth of the sun and moon when they were on the horizon. The turning points in the solar and lunar cycles, namely the solstices and the lunar standstills, would have been timeable, measureable, and noticeable.

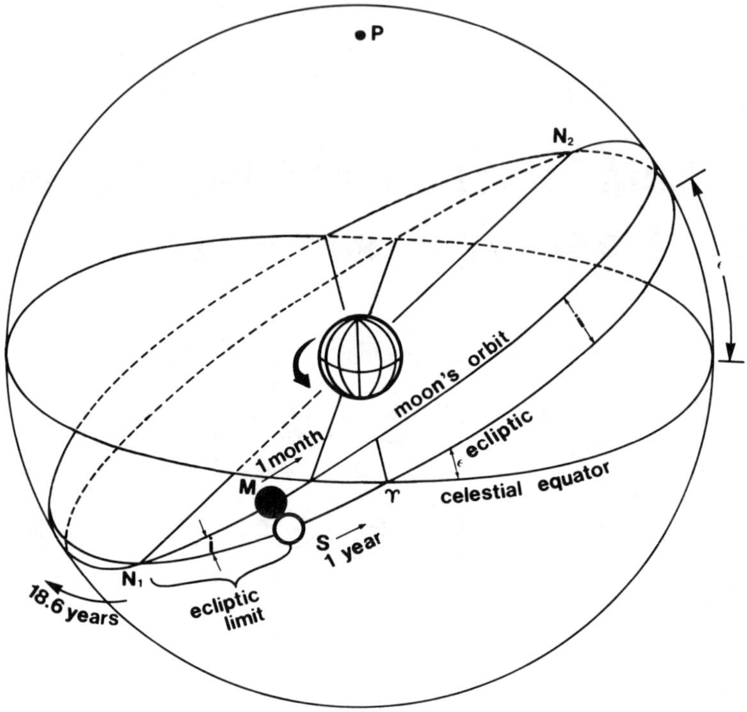

The earth occupies the center of the fictional "celestial sphere," and the earth's daily rotation makes the sky appear to turn from east to west (clockwise and toward the reader). The sun (S) and the moon (M), then, appear to follow east-to-west paths across the sky. But in a month the moon orbits the earth, and in a year the earth orbits the sun. Both of these motions make the moon and sun appear to move in the opposite sense (west-to-east, counter-clockwise, and away from the reader) from day-to-day. Also, the intersections of the ecliptic (sun's yearly path and the moon's orbit (moon's monthly path) establish a line (N_1 - N_2) that slowly turns east-to-west (clockwise). The intersections are called the nodes. The line they form is called the line of nodes, and it takes 18.6 years for the line to revolve through one complete cycle. In this diagram, i is the inclination of the moon's orbit, and ε is the obliquity of the ecliptic. The symbol γ stands for the vernal equinox. (after A. Thom, Griffith Observatory)

The astronomical declination, δ, of a heavenly body is the angular distance of the body from the celestial equator, taken at the center of the earth. As the declination of a celestial object varies, its position on the horizon, or azimuth, when it rises or sets will also vary. An observation of sunset, for example, is therefore an observation of the sun's azimuth, from which its declination can be ascertained. The obliquity of the ecliptic, ε, requires the declination of the sun to vary from $+\varepsilon$ to $-\varepsilon$ and back to $+\varepsilon$ during a year. The moon's orbit is inclined at an angle i to the ecliptic, and the places where it crosses the ecliptic are called the nodes. The moon's declination at the time of major standstill varies from $(\varepsilon + i)$ to $-(\varepsilon + i)$ and back in one lunation, or month. The nodes shift position and, in fact, rotate completely around the ecliptic in 18.6 years. This means that 9.3 years after major standstill the limits of the moon's declination in one lunation are $+(\varepsilon - i)$ and $-(\varepsilon - i)$, and this is the time of minor standstill. After another 9.3 years the limits are again $+(\varepsilon + i)$ and $-(\varepsilon + i)$.

Superimposed on the monthly motion and on the 18.6-year cycle of motion is the effect of the change in direction of the sun. This is a cyclical perturbation, a small wobble of the declination. At maximum the change is as much as 9 arcminutes, and the perturbation cycle itself lasts 173.3 days. The perturbation, Δ, affects only i, the inclination of the moon's orbit, but this shows up as a change in the moon's declination and, in turn, a change in its azimuth on the horizon.

At the major and minor standstills, when the rate of change of $(\varepsilon \pm i)$ is zero, the perturbation is most apparent. This change in the moon's declination is visible if horizon observations are taken, during the standstill season, one lunation apart, at rising or setting. Solar and lunar eclipses at major standstills can occur only at the new or full moon nearest the date when the perturbation is near its maximum (twice a year of 346.6 days, the period in which the sun completes a circuit of the ecliptic relative to a node). A trained observer would be able to predict eclipses by watching for the perturbation extreme. At the 1969 major standstill, for example, the eclipses took place when the moon was at the crests of the perturbation wave. At

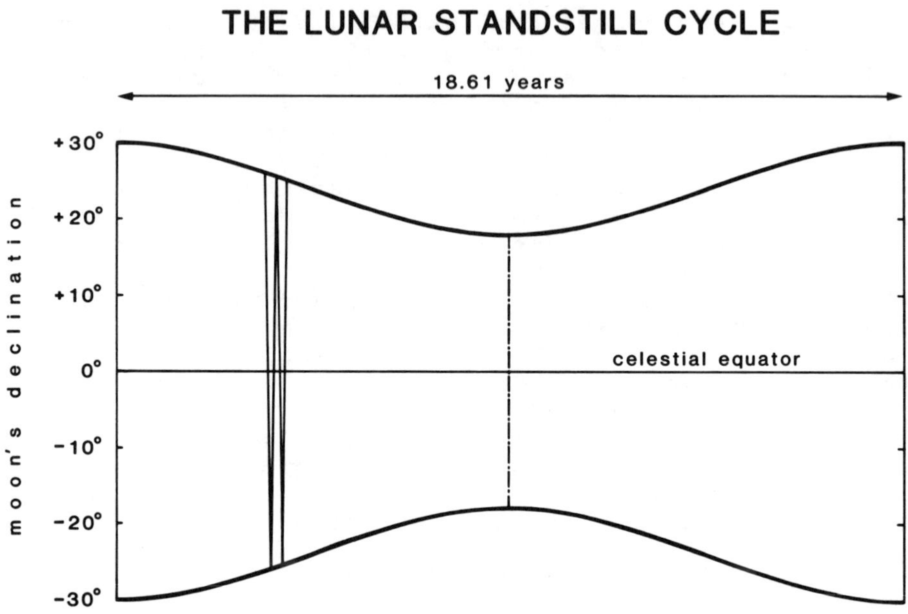

Each month the moon's declination oscillates between the extremes defined by the "hourglass" figure (from north to south and back to north again). From month-to-month, then, the moon's limiting declinations change. The time separating maximum monthly excursion to minimum monthly excursion is 9.3 years. The entire cycle, of course, lasts 18.6 years, the period of one complete regression of the line of nodes, the celestial motion responsible for this change in the moon's monthly limits. (after A. Thom, Griffith Observatory)

---

the 1978 minor standstill the ecliptic limits, and the eclipses, occurred at the troughs of the perturbation wave.

With the approach of the major standstill, observers would watch the gradual increase of each month's extreme declination. They would locate where to stand and where to look to see the moon at its greatest extreme, in the north and the south. At minor standstill, by contrast, they would watch the gradually diminishing monthly extreme. The moon's declination extremes at minor standstill are not really unique. Throughout the 18.6 year standstill cycle, the moon passes through

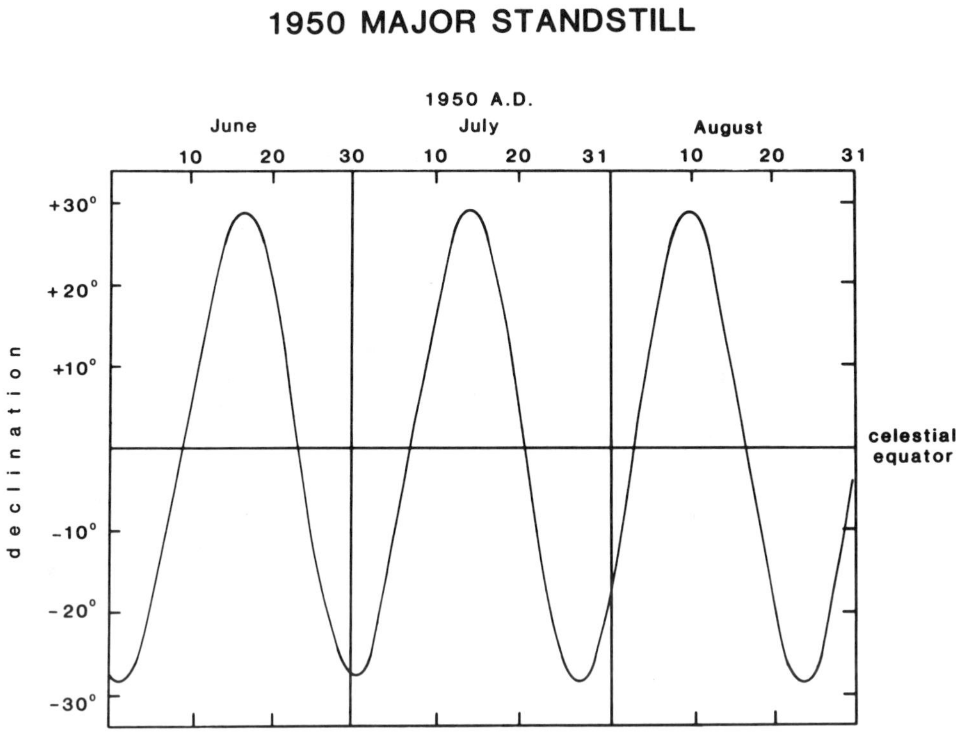

A magnified view of the monthly shift in the moon's declination over a three-month period near the time of the 1950 major standstill shows that the moon's declination maxima do not change much from month to month. This contrasts with more noticeable change at other times in the 18.6-year cycle. (after A. Thom, Griffith Observatory)

---

these declinations, above and below the celestial equator, each month. At the minor standstill, however, the perturbation cycle can be timed because the moon, then, does not go beyond these limits. And at minor standstill, it is the quarter moons, at the equinoxes, that occupy these limits.

Method of Observing

Prehistoric people constructed their observatories empirically, probably with little awareness of any of the abstract concepts of positional astronomy, except for azimuth. Undoubtedly, they could count. They had no instruments with accurate-

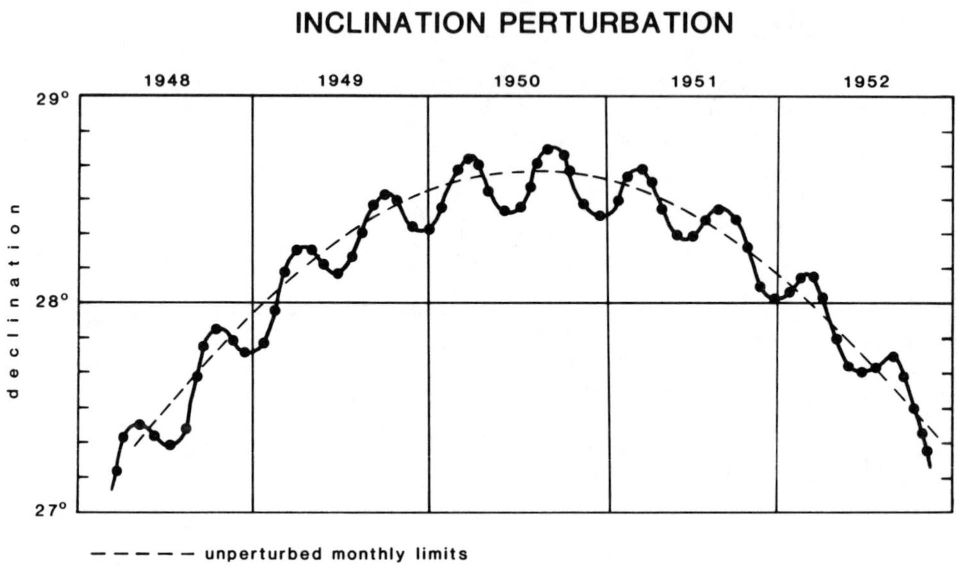

Yet one more motion of the moon, a gravitational perturbation by the sun, affects the moon's position and alters, therefore, the exact value of the monthly declination extremes. This perturbation (Δ) acts on the inclination of the moon's orbit and introduces a 173.3-day variation on declination. The monthly extremes are indicated, then, by the wiggly line superimposed upon the dotted curve, that represents the unperturbed declination limits at the time of major standstill. The perturbation therefore makes the differences between declination limits of different standstill months more distinctive. (after A. Thom, Griffith Observatory)

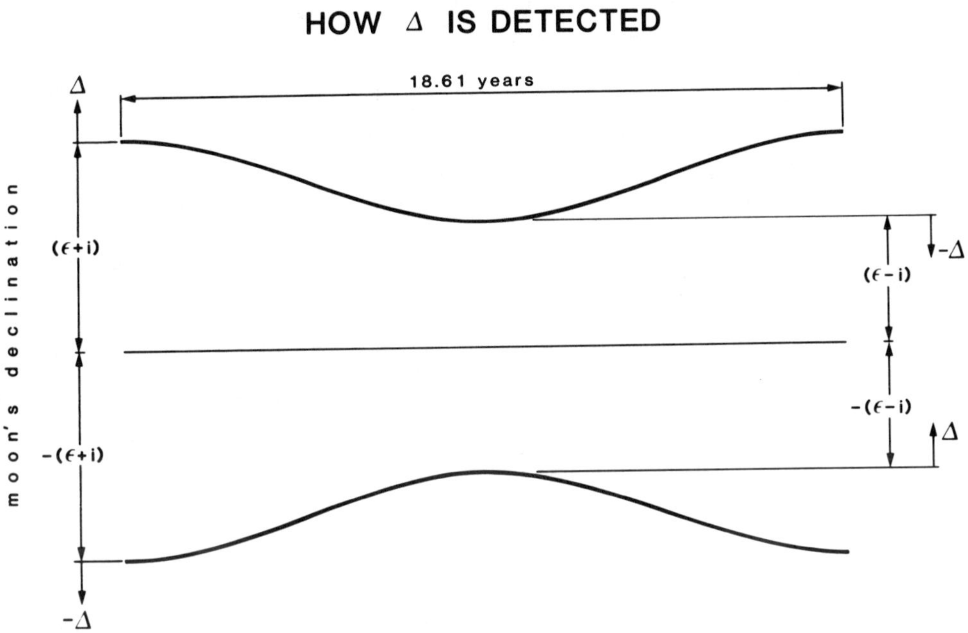

It is evident, by now, that the 173.3-day cycle of the inclination perturbation operates continuously throughout the 18.6-year cycle of the regression of the line-of-nodes. But the perturbation, Δ, can be detected only at the standstills, when the extreme moonrises and moonsets will be shifted a small amount beyond the unperturbed limits. At major standstill, the perturbation must carry the north and south moon farther north and farther south respectively. At minor standstill, 9.3 years later, the direction of the visible perturbation effect is reversed. The northernmost moon is less north, and the southernmost moon is less south. (after A.S. Thom, Griffith Observatory)

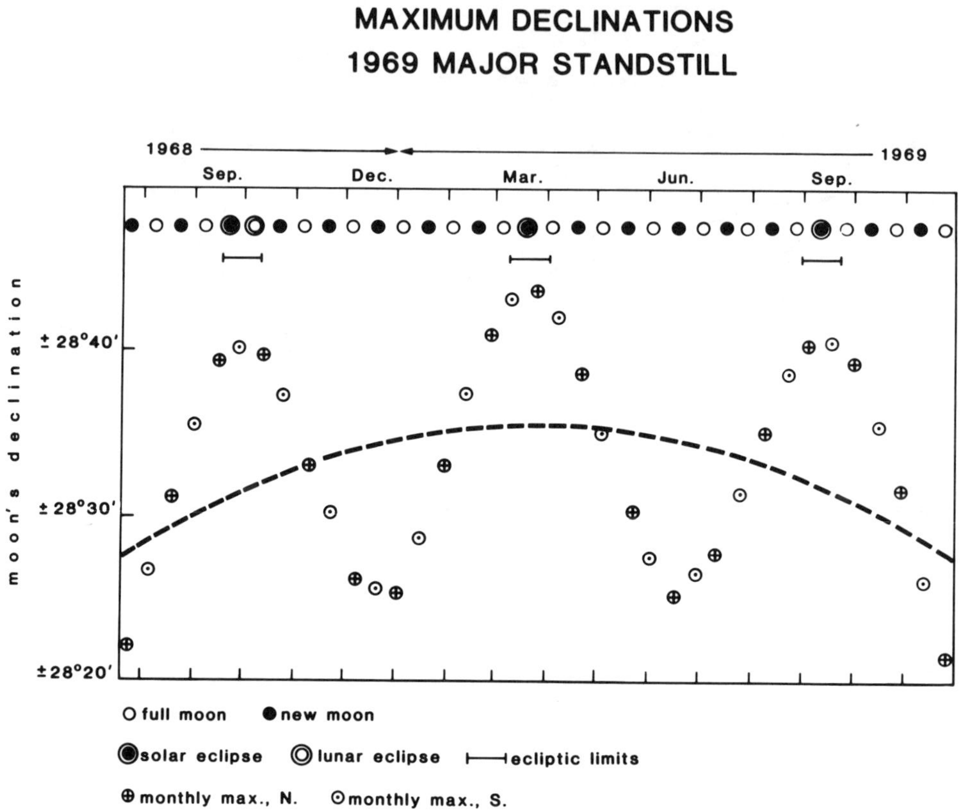

The astronomical situation at the time of the 1969 major standstill included a solar eclipse near the vernal equinox (in March), when, at the last and first quarter moons straddling the new moon of the eclipse, the inclination perturbation was greatest. Because this makes the unperturbed major standstill declinations (north and south) a little greater, the effect of the perturbation can be detected in the position of the moon. North and south declinations are combined in this diagram, which shows three cycles of the perturbation changes. Note, too, that other eclipses are correlated with the perturbation cycle. (after A. Thom, Griffith Observatory)

ly graduated circles, but they made use of the horizon's great circle. Instead of precise graduations incised on an instrument, the megalithic sky-watchers used foresights, far enough away on the horizon to permit great accuracy. The foresight was permanent, and so the observing position, the backsight, was moved to put the observer in the right place at the right time. The observer stood, for example, where the sun or the moon would appear to graze a distant foresight, perhaps a hill top. Where one must stand to see this event would be marked by a rod. By repeating this procedure each night for several nights at the time of extreme azimuth the observer collects several rods set out on a line perpendicular to the line-of-sight. By stepping back a short, constant distance each night one could even generate, on the ground, a curve which is, to some scale, a graph of the declination changing each day. The maximum point on this curve is, then, the backsight at which the extreme azimuth would be seen at the chosen foresight. The moon's turning point would not always occur, however, only at rising or setting. A method of interpolating between two evening (or morning) observations had to be developed. Two separate solar standstill sites have been discovered where the sun had actually set behind a mountain before reappearing momentarily in the notch between it and the next peak. Several sites were carefully chosen for two solar lines (each at a different calendar date), for a lunar and a solar line, or for several lunar lines.

Two Types of Observing Lines

Remains of two totally different types of lines can be extracted from the megaliths. The earlier consists of two spots, perhaps the centers, or other parts, of two circles or a circle and an outlying stone. The line joining these spots, extended to the horizon, shows the rising or setting points of the sun at an important date in the calendar, a solstice, for example, or an equinox. These lines are little more than symbolic, however. They cannot possibly give us a precise measurement of the declination involved.

A second type of line is indicated by a backsight, which may be a standing stone, and a foresight, perhaps a notch on the distant horizon. A

CONDITIONS AT THE MAJOR STANDSTILL

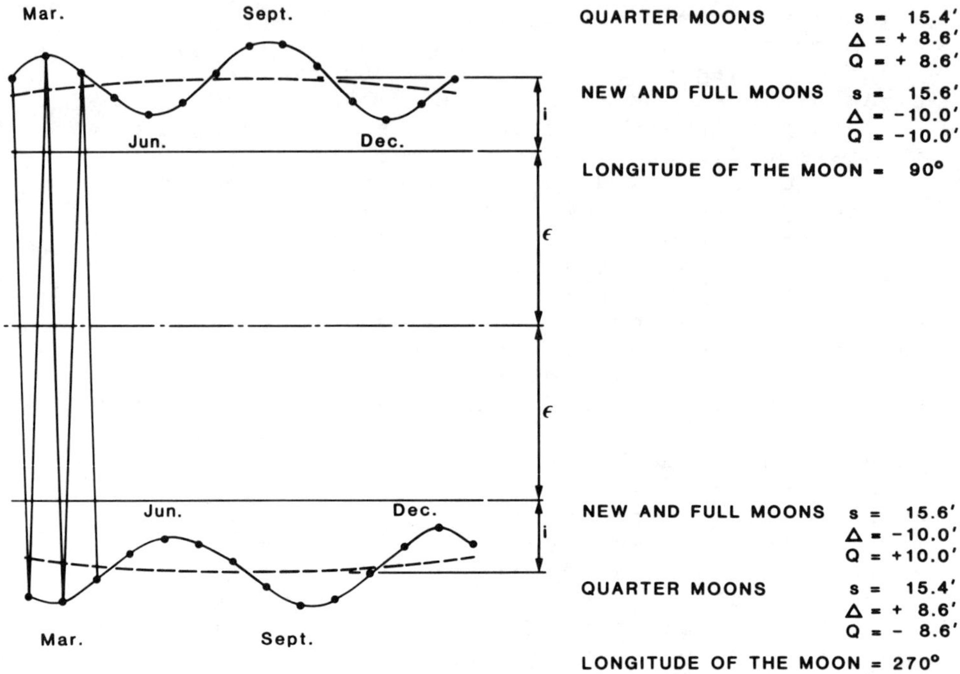

These conditions prevail for the declination of the moon at the time of major standstill. Each month the moon oscillates between the upper and lower limits appropriate for that time and achieves its greatest monthly excursion when the combination of ε, i, and Δ is maximum. At major standstill, the moon reaches maximum declination, north and south, when its longitude is 90° (north) and 270° (south) and at first and last quarter moon. Q is the expected value of β (see below), having one of the appropriate values of $\pm(\Delta \pm s)$. In this diagram, s is the moon's apparent angular radius. These declination extremes also take place around the times of vernal and autumnal equinox. (Displacement in this drawing is diagrammatic only and not to scale.) (after A. Thom, Griffith Observatory)

CONDITIONS AT THE MINOR STANDSTILL

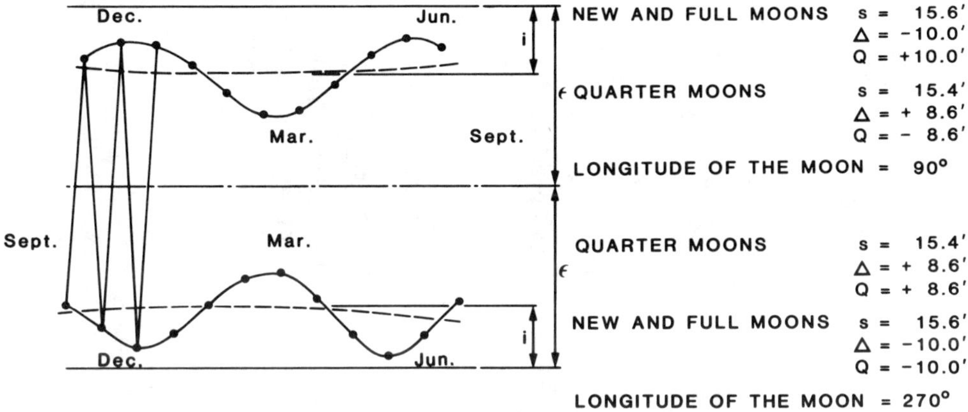

At minor standstill similar conditions prevail. The moon's monthly extremes diminish to minimum excursion at the equinoxes, at first and last quarter moon. Between major standstill and minor standstill 9.3 years elapse. (after A. Thom, Griffith Observatory)

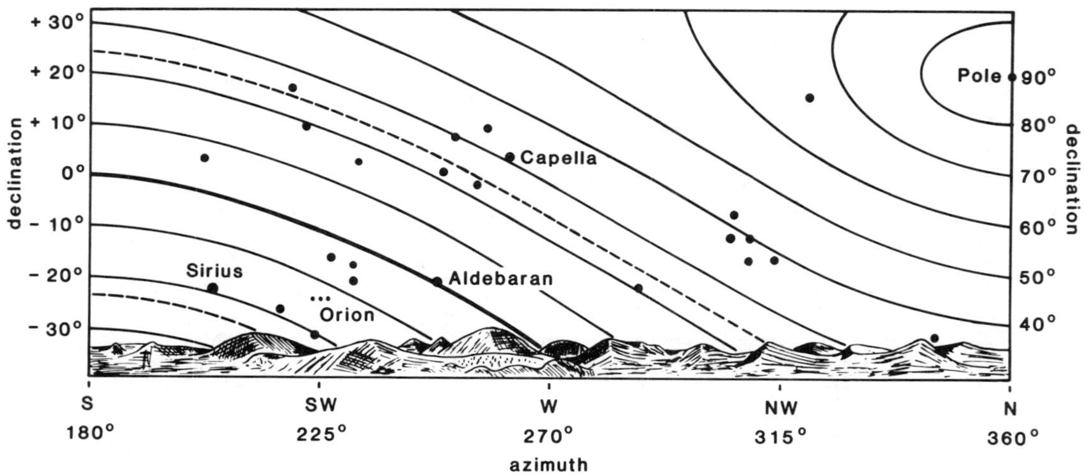

By stretching out the entire western horizon from due south (180°) to due north (360°), the paths followed by objects at various declinations can be visualized. Here the vista corresponds to latitude 55° north, and the stars occupy their positions of 2000 B.C. The broken-line curves represent the paths of the summer and winter solstice sun. Aldebaran, the bright star in Taurus the Bull, is on the celestial equator, and the north celestial pole is 55° above the horizon at the cardinal north point. (after A. Thom, Griffith Observatory)

METHOD OF OBSERVING

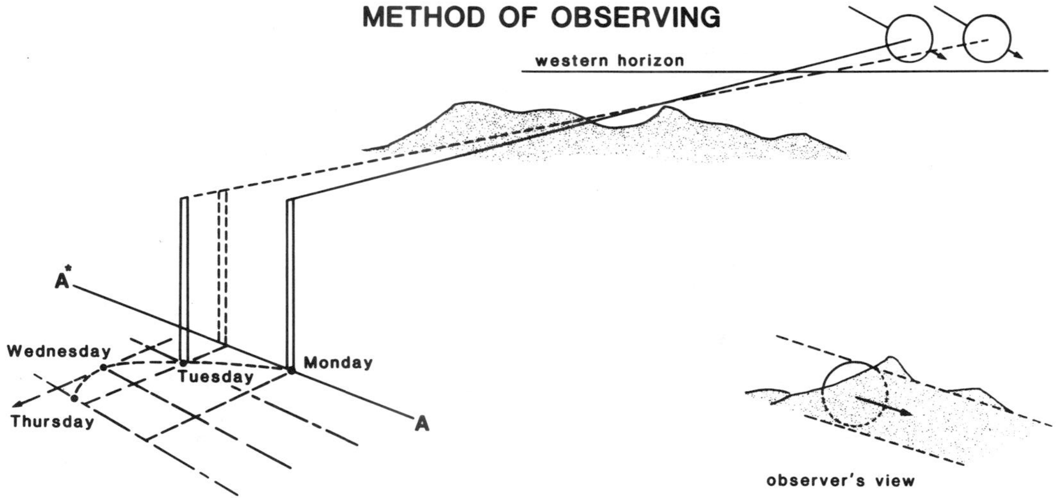

The horizon may be used to monitor and mark celestial events. The shifting position of sunset is detected by observing the sun with respect to some fixed reference, like the horizon peak in this drawing. If this represents the time of summer solstice, there is a spot where the observer must stand to see the northernmost setting sun slip behind the reference peak. The exact day of solstice is difficult to establish, but it might be ascertained by watching the sunsets on several days during the period in question. By marking the spots where he must stand to see the setting sun in line with the peak, the observer obtains a set of "places to stand." His position moves left (south) as the sun moves north and back to the right (north) as the sun returns south after the summer solstice. The southernmost rod establishes the observing point for the northernmost (summer solstice) sun. By taking one step back each night, the observer generates a parabola out of his rods' positions. The rounded end of the parabola corresponds to the summer solstice observing point for the horizon foresight in use. (after A. Thom, Griffith Observatory)

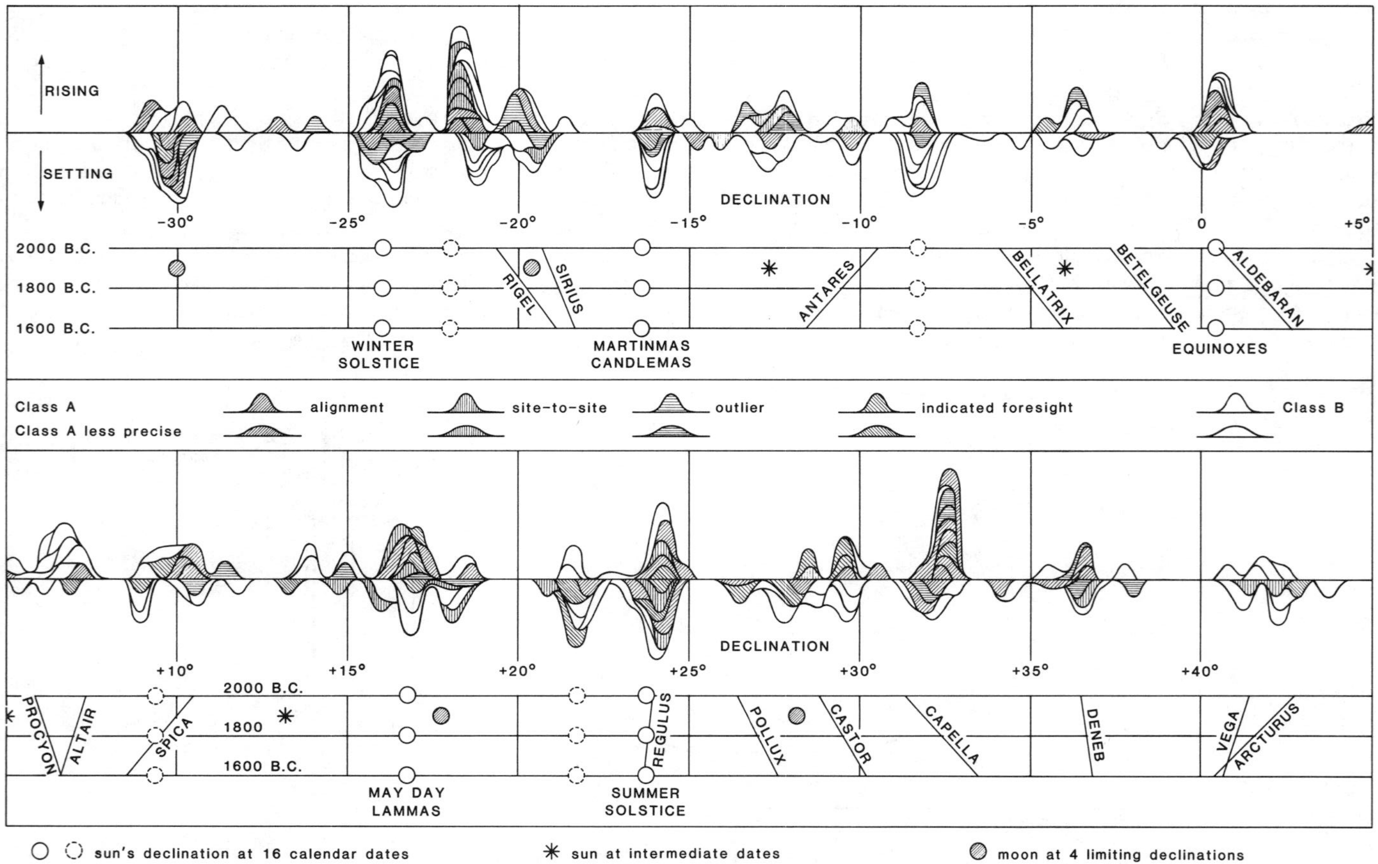
RISING
SETTING
DECLINATION
-30°
-25°
-20°
-15°
-10°
-5°
0
+5°
2000 B.C.
1800 B.C.
1600 B.C.
RIGEL
SIRIUS
ANTARES
BELLATRIX
BETELGEUSE
ALDEBARAN
WINTER SOLSTICE
MARTINMAS CANDLEMAS
EQUINOXES
Class A
Class A less precise
alignment
site-to-site
outlier
indicated foresight
Class B
DECLINATION
+10°
+15°
+20°
+25°
+30°
+35°
+40°
2000 B.C.
1800
1600 B.C.
PROCYON
ALTAIR
SPICA
REGULUS
POLLUX
CASTOR
CAPELLA
DENEB
VEGA
ARCTURUS
MAY DAY LAMMAS
SUMMER SOLSTICE
sun's declination at 16 calendar dates
sun at intermediate dates
moon at 4 limiting declinations

Megalithic alignments may correspond, at least in some cases, to astronomically important directions on the horizon. By transforming the observed azimuths to declinations and by analyzing these "observed" declinations collectively, we may evaluate the astronomical significance of these alignments. Key dates in the annual solar cycle are indicated by open circles (winter solstice, Martinmas/Candlemas, equinoxes, May Day/Lammas, and summer solstice), and these subdivide the year into eight parts. Additional dates that permit a 16-interval calendar are shown with broken-line open circles. Still more intermediate dates, for further subdivision of the year, are indicated by asterisks. Filled circles mark the standstill declinations of the moon. Declinations of stars vary because of precession, but the lines provided here display the changes for several bright stars during the period 2000-1600 B.C. If each alignment is represented as little gaussian "hill," the height of which increases with greater precision of alignment (Class B alignments are those with less identifiable foresights), many such "hills" for the same astronomical target will build into a taller hill at the appropriate declination. These peaks are seen to occur at a number of important astronomical directions. (after A. Thom, Griffith Observatory) precise declination of a limb of the sun or the moon may be obtained, but, such sightlines were not used until about 2,000 B.C. We shall confine our interest here to only these precision alignments. A suitably distant natural foresight can permit better than one arcminute precision in declination.

Possibly some prehistoric observatories also incorporated artificial foresights, constructed of perishable material. This may have happened at Stonehenge, but we do not know for certain what such foresights may have looked like. We can guess that they might have been huge mounds of turf or brushwood. For such markers to be clearly visible to the unaided eye when silhouetted against the moon's disc, they have to subtend an angle of over one arcminute. This is equivalent to about a foot and a half at one mile, distance. If we assume the foresight was 10 miles away, we require a target about 15 feet square, quite a reasonable size.

We search for astronomical lines by standing at a marked backsight. There we carefully measure the horizon profile. If it has a feature that marks a significant declination, we assume we are at a real observing point. An observing point, or backsight, may be a large stone, sometimes particularly large as at Ballinaby in Islay. If the stone is a slab, the long side will point to the foresight. (The Comet Stone at Brogar indicates, unequivocally, the Mid Hill notch.) At another site we may find two or more stones set on a line that points to a foresight (at Dunskeig, for example). In Orkney, at Brogar, there are earth mounds which serve the same purpose. A foresight can be used for either the upper or lower limb, where upper and lower refer to declination and not to altitude.

When prehistoric people wanted to mark the setting point of the moon's center, a more complex technique was needed. One person would find the right spot to see the upper limb coincide with the foresight. A second person would locate the right place for the lower limb. Midway between the two positions (on relatively even ground) is the sighting position for the center of the orb. In a few instances this spot is marked by a stone.

For all cases, as the moon slipped behind the

foresight, the observer moved about until an exact position was found. The observer then marked the position with a stake. After doing this on consecutive nights, month after month, at each standstill the megalithic astronomer deduced the final position from the stakes on the ground. If, in later cycles, an observation were missed due to bad weather, its position would be marked in stone for permanence.

Ideal Foresights: Kintraw

At Kintraw, in western Scotland, we find what may be an ideal foresight for the sun. There, at the winter solstice, the setting sun disappears behind a hill before reappearing momentarily in a notch. Before finally vanishing behind the peaks of Jura the upper limb twinkled in the V between the mountains Ben Shiantaidh and Ben a' Chaolais. The effect is quite dramatic. As the winter solstice approaches, the observer finds it necessary at each successive sunset to move to the right along the platform by progressively a smaller amount, until the actual evening of the solstice. Dr. Euan MacKie (1975) has described this site fully and reported that only to the left of the backsight, which is sited on a track or pathway, was the platform leveled with small stones. Here the archaeological evidence agrees with the astronomical requirements. I know of three other archaeologists who saw the cobbled surface of paving stones during the excavation. In addition to Dr. Bibby, who did the soil mechanics work for Dr. MacKie, these three also held the opinion that the paving was manmade.

Many visitors to Kintraw have been very unlucky and because of weather failed to see the Jura notch, but I have watched the sun set in the notch while standing on the stone on the platform three days before the winter solstice. No terrestrial refraction could ever make the notch, which has an altitude of 25 arcminutes, disappear below the near foreground under any weather conditions experienced in the Inner Hebrides.

The site at Kintraw has, naturally, attracted many people, and here I quote Alexander Thom in some detail.

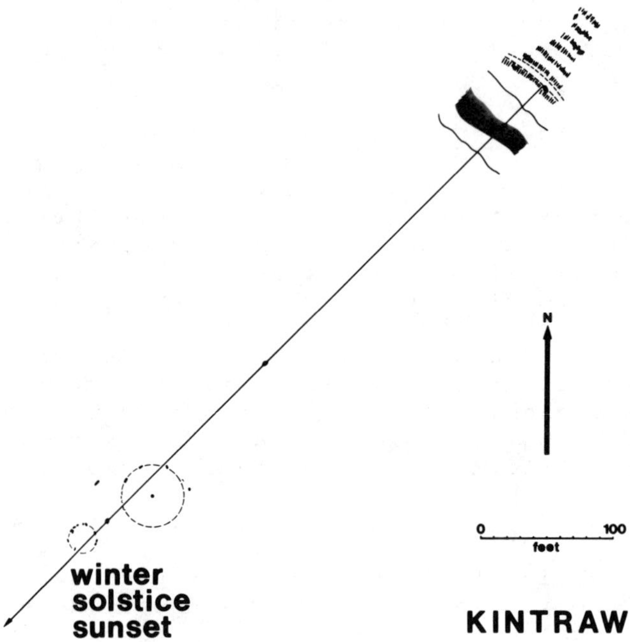

The large cairn and standing stone at Kintraw are on the winter solstice sunset line-of-sight that extends from the "platform," on the steep hill to the northeast, to the Paps of Jura, an island about 28 miles to the southwest. (after A. Thom, Griffith Observatory)

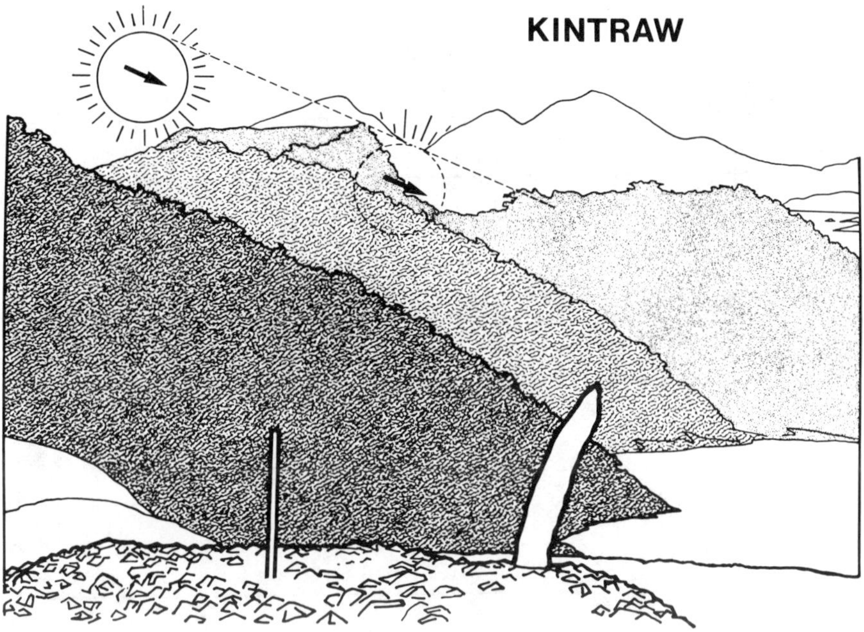

From the Kintraw stone "platform," with its boulder-notch, the winter solstice sun appears to sink behind the first peak on distant Jura, and then its upper limb reappears in the V of the horizon and gleams momentarily with a parting shot of sunlight. (after Euan MacKie, Griffith Observatory)

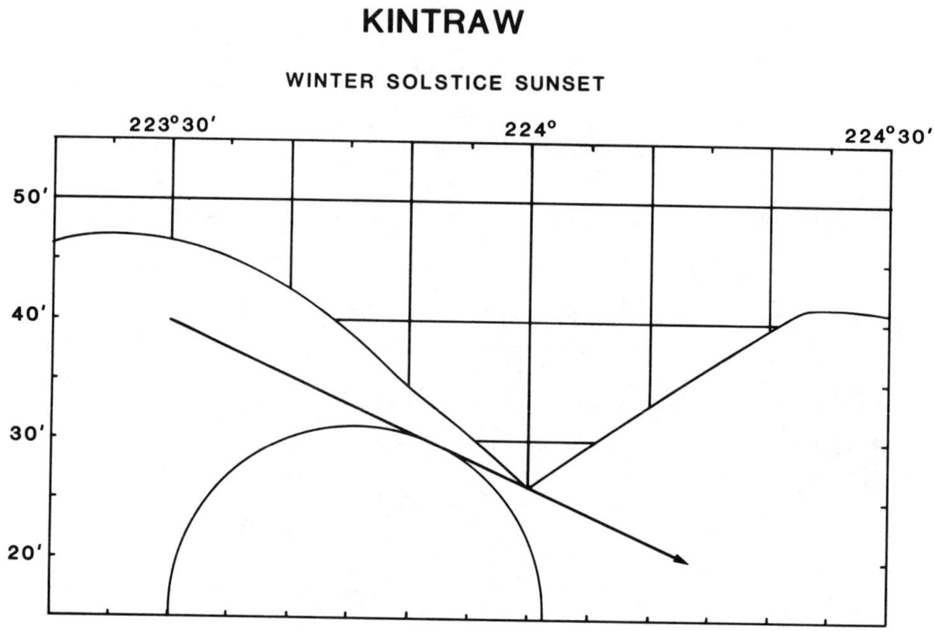

This magnified view of the horizon profile for the Kintraw winter solstice sunset shows the fine tuning of this potential solar observatory. The peak on the left is Ben Shiantaidh; that on the right is Ben á Chaolais. A conspicuous last beam from the sun's upper limb emerges from the notch between them. (after A. Thom, Griffith Observatory)

The profile of the mountains given in Fig. 12.1 (*M.S.B.*) was constructed very carefully from the map contours, but it has turned out to be almost exact. My difficulty at that time was that standing at ground level the notch is hidden by a near-by ridge. I pointed out that if the cairn were much larger it would be possible to see the notch when standing on the top. There are remains of large stone fanks (sheep corrals) nearby and it is almost certain that these were built out of the cairn.

On my next visit I was accompanied by my grand-daughter, Susan. In order to get a clear view we decided to climb up the slope on the north side of the gorge. We eventually got the theodolite over the gorge by going upstream. The difficulty here was the very high barbed wire fences, but this route would have been the natural approach at an earlier date when of course there was no road bridge over the stream at Kintraw. And so we came to the track along the hill-side *from the east.* I immediately moved along the track until I got the menhir in line with the notch. I then set up the instrument and asked Susan to get ready to book. As I was levelling the instrument, I drew Susan's attention to the fact that there was another stone on the line on the plateau, S_3 on Fig. 4.3 *M.L.O.* She immediately replied "And look what I am sitting on." That is how stones S_3 and S_4, what I call "The Susan Stone", were found. From this point on the track there is a perfectly clear view of the hills at the notch. I carefully measured the altitude and azimuth of the notch with a good theodolite used correctly, Table 4.1, *M.L.O.* Later on Robert L. Merritt found stone S_6 higher up the hill placed there obviously for a watcher to give warning that the sun was about to appear. Notice that this stone is higher in altitude and is slightly to the right of the sight line so that a man placed here is in a position to give a few seconds warning

> that the sun is going to appear. We have plenty of witnesses that this stone was there but how long it will be allowed to remain is another matter.
>
> Note that the line S_3-S_2 indicates Choire Dubh Bheinn and so gives a declination $-(\varepsilon-i-\Delta-s)$, Table 7.1, *M.L.O.* We cannot of course substantiate the suggestion in Fig 12.1 in *M.S.B.* that point B gave a calendar date, but it is possible.

Four types of foresight used by the megalith builders have been classified (Thom & Thom, 1980). All are capable of defining the position of the moon's limb with a precision better than one minute of arc. To do so, the backsight must be on a reasonably flat area of ground, and finding a position complete with a suitable foresight and backsight cannot have been particularly easy. The task was made more difficult by the evident desire to tie two or more foresights into the same backsight, or at least to have several backsights grouped together. Examples of this are the lunar site at Brogar in Orkney and Temple Wood in Argyll or the solar site at Ballochroy on Scotland's Kintyre peninsula.

Accurate Surveying Is Necessary

It soon became apparent that the lunar observatories may have been set up with much greater accuracy and precision than at first had been thought possible. Accordingly, a survey of a possible megalithic observatory requires a precise and accurate record of the azimuth and altitude of each foresight, as seen from the observer's position at the backsight. After correcting an observed altitude for refraction and parallax, the declination of the line is evaluated. The azimuth of the foresight is usually obtained by taking the mean of three time-azimuth observations of the sun using a good theodolite and chronometer. From readings of the horizon its profile is plotted. The chronometer is checked daily by radio, to within ½ second of time, and the position of the instrument is carefully recorded along with a reference object, usually some easily recognized feature in the distant landscape. Darkened eyepieces are used while observing, first, the leading and, second, the trailing edge of the sun. It of course is necessary to maintain the theodolite in good condition. Accuracy of its vertical circle is checked frequently.

The Solar Calendar

When my father began to survey megalithic sites, he surveyed lines of three stones or two slabs in line or outliers from circles. From the observations obtained he plotted a histogram and represented each alignment as a small gaussian curve. The technique for presentation of observed declinations, risings and settings, has evolved over the years that site surveys have continued. The subject has grown and expanded, and refinements have been made.

Among observed alignments are declinations that correspond to the sun's position on the dates of May Day/Lammas and Candlemas/Martinmas. These, along with the solstices and equinoxes, divide the year into eight parts. There are also concentrations of alignments at declinations midway between these, and the result is a 16-part year. Had the lines been set out at random, the histogram of these declinations would have looked very different. The clear grouping of alignments on key declinations cannot be explained by chance. We infer, therefore, that the megalith builders used a solar calendar with a 16-part year. An up-to-date paper on the solar calendar was presented in Oxford (September 1981). It further strengthens the case for the 16-part year.

Because the earth's orbit around the sun is elliptical, the observed positions of the equinoxes differ from due east-west by about half a degree, evidenced on the histogram(*M.S.B.*). This results from determining an equinoctial sunrise or sunset by counting days and dividing the year into equal halves. Astronomically, however, the equinox is defined by the intersection of the equator and the ecliptic. An empirical, day-count approach to the equinoxes shifts the alignment from east-west by the amount and in the direction of the observed declinations. Although we cannot review here all of the data available on solar alignments, Kintraw may still be regarded as a primary example of a site at which precise solar astronomy was possible.

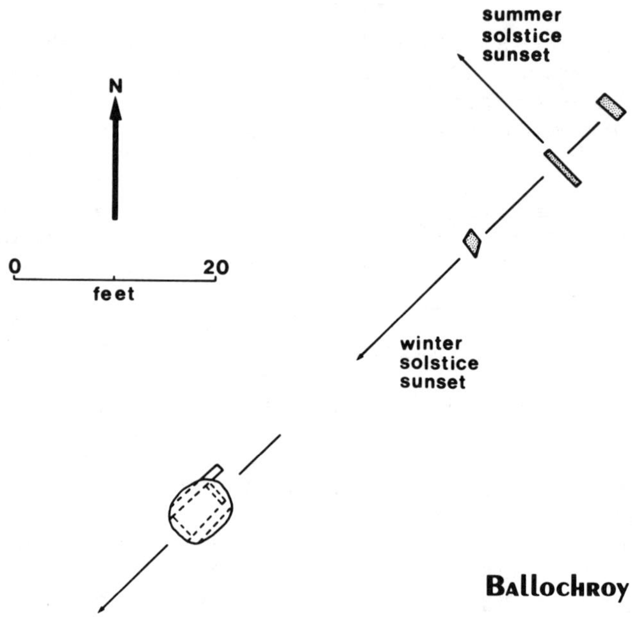

A stone chamber, or kist, and three standing stones at Ballochroy establish a line oriented to the southwest and winter solstice sunset. The large central upright slab itself appears to be oriented toward the direction of summer solstice sunset, in the northwest. (after A. Thom, Griffith Observatory)

There, we believe, the midwinter sunset was observed from the platform or ledge constructed specially for this purpose. Part of the ledge was carefully excavated by Euan W. MacKie and his results are consistent with the hypothesized astronomical use.

Twenty-nine miles to the south of Kintraw is the site at Ballochroy, where a line of three remaining stones and a kist (small stone chamber) points to the position of the midwinter sunset on the rocks at the end of Cara. Standing on the line shown by the northwest edge of the three stones, I have had no difficulty in seeing the fall of Cara Island (*M.S.B.* Fig. 12.2), which indicates from this part of the site the upper limb of the sun setting at winter solstice. Recently planted trees are now beginning to obscure Cara.

It has been pointed out by archaeologists that a mound of stones was covering the cist on this line at one time. There is no mound there now. The line of stones indicates the solstitial sunset. Was the mound there before or after the erection of the menhirs? It may have been low enough for observers to see over its top. They may have stood on its top or stood aside. A side movement of 6 feet to left or right would give an azimuth change of about half a minute. When Alexander Thom first surveyed this site just before World War II, his survey unfortunately was interrupted by a very heavy shower of rain. His rather blurred notebook shows that there were other stones to the east of the main stones. There is no trace of them to be found on the ground today. Recent inquiries of the locals were unrewarding. The stones could have been backsights for the foresight at the fall of Cara. As the sun approached the winter solstice position, the observers would move in from the southeast a little farther each evening until the solstice. Did it matter if the last positions were behind the cairn? There was plenty of room at the site to move about to finally find the correct spot.

The outline of Cara also might have been affected by the height of the tide or by mean sea level. A change in level of 12 feet would affect the profile by about one minute of arc. The fact remains that winter solstice does show up from this part of the site.

BALLOCHROY, KINTYRE

LOOKING SOUTHWEST

The view along the line of three stones at Ballochroy carries the eye to the kist and to the island of Cara, about 7½ miles away. (after A. Thom, Griffith Observatory)

WINTER SOLSTICE SUNSET OVER CARA FROM BALLOCHROY

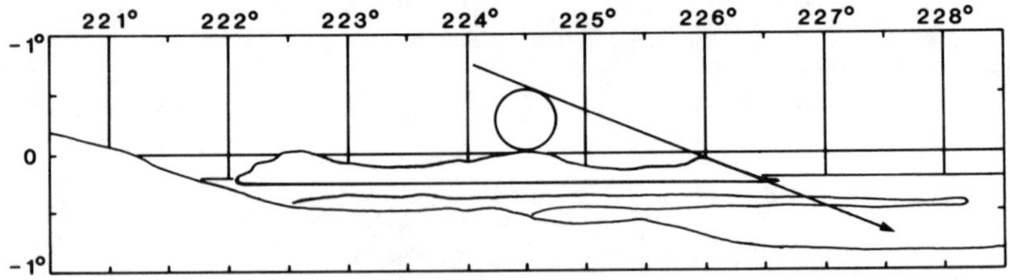

From Ballochroy, the island of Cara acts as a foresight for the winter solstice sunset. (after A. Thom, Griffith Observatory)

Of equal importance to the above line is the orientation of the slab, the center one, which unambiguously indicates Ben Corra, on Jura. There the horizon is slightly steeper than the slope of the sun's path. About three minutes after the sun sets behind Ben Corra, it reappears momentarily... provided the observer occupies the spot marked by the stones of the site. This site is more fully described in the literature and is important because it includes alignments for both winter *and* summer solstice.

Change in the sun's declination is very small at the time of the solstices. To determine the date to a day's precision requires careful observations and sites like Ballochroy or Kintraw. There are no other points in the year as obvious as these turning points, and so Ballochroy and Kintraw are extremely important. It has to be pointed out here that one observation on the day of the solstice, if missed, could be easily replaced by halving the number of days between observations made at a fixed point near to the final backsight for recording the solstice. The first observation could be made, say, 5 days before, and the second, 5 days after solstice. Relatively flat ground is necessary, or at least there has to be freedom to move along a straight line passing through the final backsight. The remaining calendar dates throughout the year do not require such accuracy because the sun's declination is then changing rapidly--as much as 21 arc-minutes per day. The year's remaining 14 calendar dates do not require such precision because the sun's declination is changing rapidly at those times. A day's movement on the horizon is quite obvious.

Lunar Alignments

Groupings of alignments also appear at important lunar declinations. In 1967, the lines near to the moon's four standstill declinations were analyzed. Observed declinations (δ_o) in good agreement $\pm(0^{\circ}.8-1^{\circ}.0)$ with the *expected* values (δ_e) were examined, and differences ($\beta = \delta_o - \delta_\varepsilon$) were plotted for each line. The way in which the gaussian representations of these alignment differences form a peak is remarkable. They group at differences that correspond to the two limbs of the moon. This is exactly what should be measured if prehis-

SUMMER SOLSTICE SUNSET
OVER CORRA BEINN, JURA
FROM BALLOCHROY

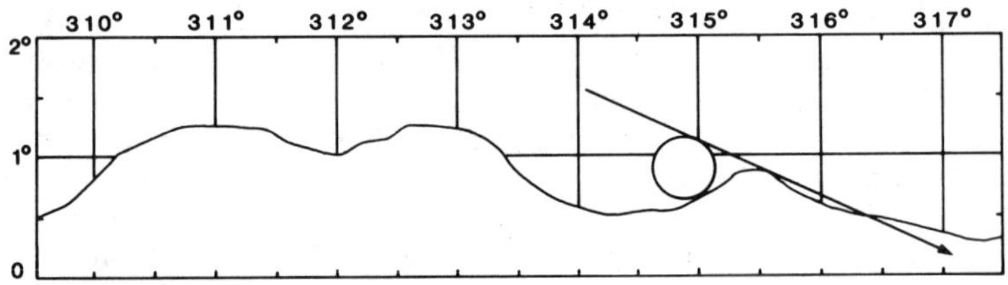

A precise observation of summer solstice sunset also seems to be possible from Ballochroy. The sun slips behind a conspicuous foresight, Ben Corra (the northernmost Pap of Jura). The peak is about 19 miles away, and on the solstice, the limb just grazes down the northern slope. Before and after the solstice, the upper limb will travel invisibly once it sets behind the top of the peak. (after A. Thom, Griffith Observatory)

DIFFERENCE BETWEEN OBSERVED DECLINATION AND EXPECTED LUNAR DECLINATIONS

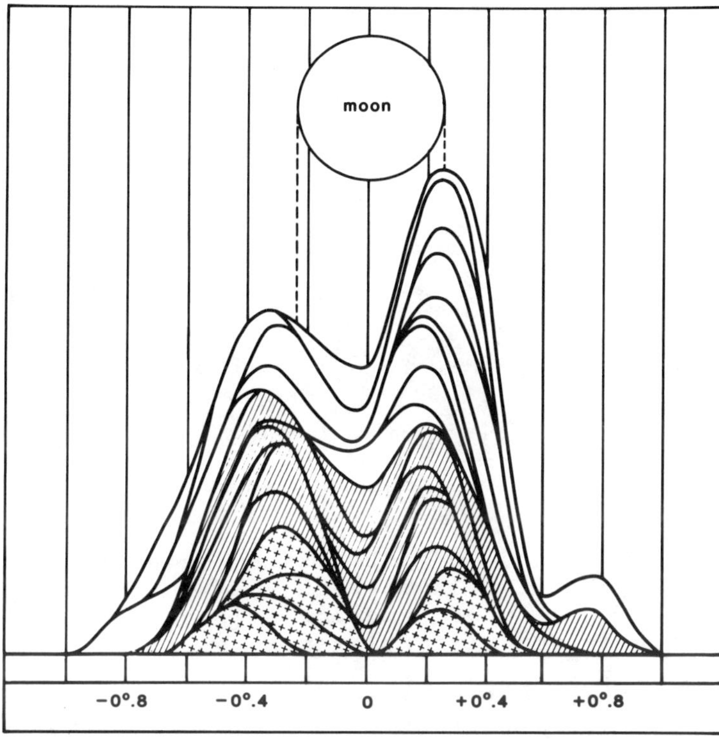

We suspect that the lunar declinations in our 42 megalithic alignments represent genuine attempts to mark precisely the observed position of the moon. When expected declinations, $\pm$ $(\varepsilon \pm i)$ are subtracted from observed declinations and collected together, the residual differences, β, combine to form a two-peak "mountain" centered on the moon's center. The two peaks correspond to the upper and lower limbs of the moon, and these are exactly what must be observed to establish a precise knowledge of the moon's behavior. The most precise alignments are textured with + signs. Those next in reliability are shaded. The least reliable gaussians are unshaded. (after A. Thom, Griffith Observatory)

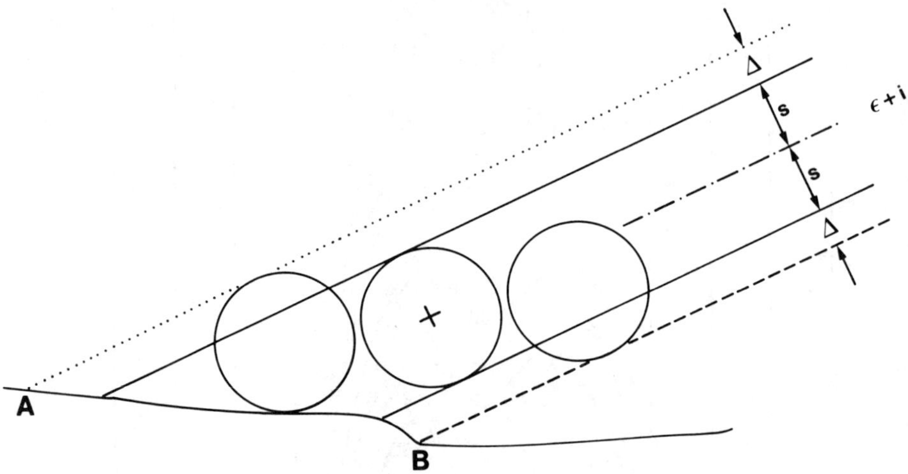

The declination zone of interest to megalithic moon watchers we call the lunar band, and it establishes the limits of the upper and lower limbs for the declination of any particular standstill. This example shows the major standstill northern moonrise. The dotted line indicates the path of the upper limb at maximum perturbation. The broken line shows the path of the lower limb at minimum perturbation. Both limbs of the unperturbed moon follow the solid lines. Points A and B mark the limits on the horizon that correspond to declination limits of the lunar band. (after A. Thom and A.S. Thom, Griffith Observatory)

toric observations were made of the limbs. Practical observation would use the limbs. This double-peak pattern appeared in the data quite naturally, and it is additional evidence in favor of the lunar significance of the lines. This was the first clear indication that the moon had been observed in a precise and practical way.

With the realization that the megalith builders had had a reliable calendar, A. Thom began to consider the possibility of their predicting future astronomical events, eclipses, for example. Alignments on the lunar standstill declinations were known. Were the megalithic observers aware of the small wobble, or perturbation ($\pm\Delta$) in the moon's orbit? To investigate this possibility, accurate surveys of alignments were needed, and we now survey to within 1 arcminute in azimuth and 20 arcseconds in altitude. The perturbation modifies the monthly lunar extremes continuously, but it can only show up at the standstills. We are only interested, therefore, in the bands of lunar declination that straddle the standstills and are defined by the amplitude of the perturbation and the angular diameter of the moon. When we visit a site we look at the places on the horizon which correspond to these declinations. There are eight possible declination zones, and we have to measure the altitude and azimuth of the horizon segments that correspond to those zones. We plot the horizon profile, and if there is a notch, mark, or other clearly defined foresight of any kind on the profile, its declination is calculated. If this declination turns out to be pertinent, we judge we are at a megalithic backsight.

More Lunar Lines

In recent years many more lunar observing sites have been surveyed. A careful analysis of 42 of these lines provides an even clearer demonstration of the precision the alignments may contain. If we represent the difference between observed declination and expected declination for each line as a gaussian profile, the combined data suggest alignments on the top or bottom of the perturbation cycle were favored. Our own knowledge of the history of these investigations (since 1967) convinces us that the recent analysis reveals real intent in the alignment of megaliths. Others are not convinced. They maintain that the way in which we identify foresights is not objective. Accordingly we set up strict terms of reference for the selection of certain sites from our collection: (a) The backsight marker must give some indication of the direction in which to look to find the foresight. (b) The foresight must be inside of the appropriate lunar declination band. (c) There must be only one point which could be used to indicate the declination. The use of these strict criteria cut down the number of sites to 20, but from these alone we found a probability level of about 0.2 percent.

Over the years our methods have evolved as our knowledge has increased. By 1971, about 80 possible lunar alignments were known. Because this subject is continually developing, our ideas change. A number of possible alignments have been rejected for various reasons. Since 1971 new ones have been found. We have summarized some of the rejected alignments in the appendices.

By establishing the likelihood that 20 lines indicated lunar positions observed in prehistoric times, we feel we may relax the restrictions at this point and examine the 42 most reliable lines in more detail. In this way we hope to find out just how the megalith builders used their observatories.

In our analysis of the 42 best lines, the declination of the actual alignment was again calculated from altitude and azimuth, both measured by theodolite. Details of such a calculation follow. Refraction for the appropriate time of day was applied, as were mean parallax values (56.4 arcminutes at the equinox and 57.4 arcminutes at the solstices). From each line a value for the obliquity of the ecliptic (ε) can be calculated, if desired. From these an attempt at site dating may be made.

Additional corrections and refinements appear in our calculations. *Curvature* is one of these. Another is *graze*. Graze is the extra bend in the path of a ray of light that passes over a ridge, especially in the early hours of darkness when the ground may cool more quickly than the air. The graze correction in these calculations is about

-0.6 arcminutes and is applied to the refraction. It can be shown that the graze correction itself should be corrected, but the amount is small and the effect is negligible.

A simple calculation of the declination δ_o indicated by an alignment and of the value of obliquity of ecliptic, ε, that results follows.

The following data are needed, for any site considered and the lunar configuration appropriate to it.

1. height above sea level (for mean barometric pressure)
2. latitude and longitude
3. azimuth of alignment
4. observed altitude of horizon in vicinity of foresight
5. relevance of the equinoxes or of either solstice
6. approximate hour angle of moonrise (or moonset), measured from the south
7. celestial longitude of the moon
8. celestial longitude of the sun
9. air temperature
10. refraction angle
11. graze
12. parallax
13. geocentric altitude
14. Δ, the relevant perturbation
15. s, the relevant semi-diameter
16. i, mean value of the inclination of the moon's orbit, $5^{\circ}8'.7$

In the calculation the celestial longitudes of the sun and moon are measured anti-clockwise from the First Point of Aries. The time elapsed since midnight is the hour angle of moonrise or moonset minus the longitude of the moon. In this example, it is 90 degrees. We chose Ballinaby, Islay (the upper foresight), as our alignment.

height	150 feet above sea level	
latitude	$55^\circ 49.0'$	
azimuth	$327^\circ 28.0'$	
altitude	$0^\circ 44.2'$	
nominal declination	$\varepsilon + i + \Delta + s$	
month	March	and September
hour angle	143°	143°
longitude of moon	90°	90°
longitude of sun	0°	180°
hour (hours elapsed since midnight)	3.5	15.5
temperature (degrees Fahrenheit)	39	54
refraction (r)	-27.4	-26.3'
graze (g)	-2.0	-0.0'
parallax (p)	57.2	57.2'
geocentric altitude (hg)	$1^\circ 12.0$	$1^\circ 15.1'$
declination	$29^\circ 24.0$	$29^\circ 26.9'$
mean declination δ_o	$29^\circ 25.4'$	
perturbation	7.1'	7.1'
semidiameter (s)	15.6'	15.6'
inclination (i)	$5^\circ 08.7'$	
obliquity	$23^\circ 52.6'$	$23^\circ 55.5'$
mean obliquity	$23^\circ 54.0'$	

The refraction (r) is from tables and a diagram in *The Nautical Almanac* and is based upon barometric pressure (mean value for the height) and temperature. Graze (g) is taken as -2 arcminutes from one hour before sunset to one hour after sunrise. It is zero during the remainder of the day. Geocentric altitude is the sum of parallax, refraction, graze, and altitude. Here, declination is given by

$$\delta = \arcsin(\sin \phi \sin h_g + \cos \phi \cos h_g \cos \text{azimuth})$$

This calculation was made in 1978. We now use, for the equinoxes, p = 56.4' and s = 15.4', and instead of Δ = 7.1', we use 8.6', as explained. We have no way of telling which--March or September--conditions are appropriate. Therefore, we have taken the mean. This assumes that observers could have watched a waning "half moon" in daytime at 15:30 hours (about 3:30 p.m.) in September. Would this have been possible at low altitude, were the sky clear? The waxing "half moon" could have been seen, of course, clearly in the darkness at 3:30 hours (3:30 a.m.) in March.

When we examined the 42 differences between measured declination and expected declination, we again found clustering around the expected values, that is, those that correspond to the limb of the moon at the upper and lower limits, and in the center, of the perturbation cycle. We represent each difference with a gaussian profile, and a larger gaussian "hill" or data node, builds up where several alignments on the same declination agree. S.R. Broadbent has developed a statistical method for dealing with such a set of uniformly spaced data nodes. The spacing in this case is about 8 arcminutes. We can, with Broadbent's method, find the probability that the distribution of nodes (differences between expected and measured declinations) is random. In this case, the probability level is very low: 0.01 percent or about one chance in ten thousand. This result is very low and suggests the data nodes are not random at all. Instead we conclude they represent the real targets of megalithic lunar alignments.

Using the Old Stones

It is easiest, perhaps, to understand how we believe the megalithic lunar observatories were

HOW THE ACTUAL DECLINATIONS (δ_o) OF LUNAR LINES DIFFER FROM THE EXPECTED DECLINATIONS ($\delta_e = \pm(\epsilon \pm i)$) FOR THE STANDSTILL MOON.

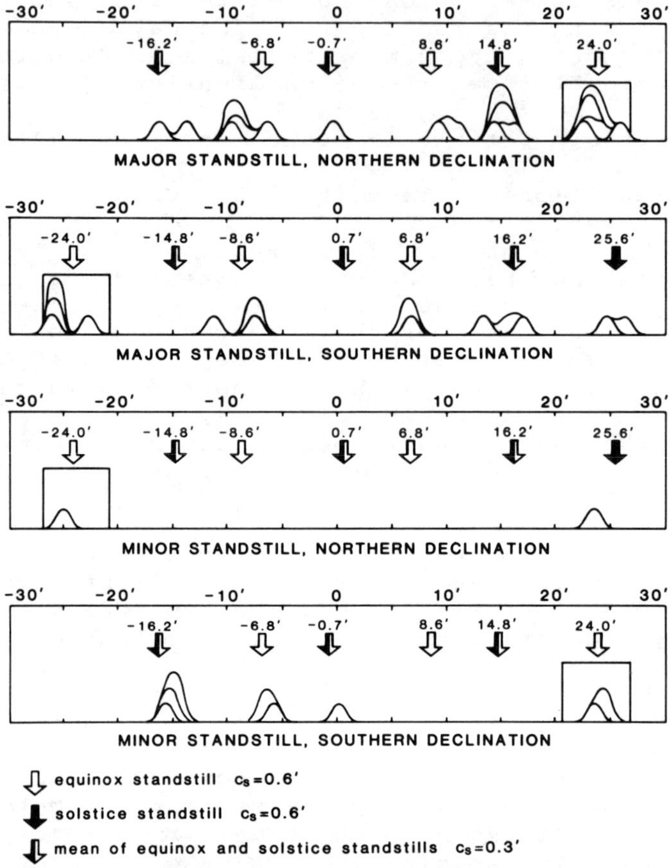

Here again gaussian hills represent individual alignments, suspected to be lunar. When the expected declinations (which include the appropriate combinations of obliquity, ε, and inclination, i) are subtracted from the observed declinations of the alignments, the differences--or residuals--should represent the various combinations of perturbation and displacement of the limb(s) that are relevant for horizon astronomy. Calculated values for these residuals are called Q, and their values assume

$$\Delta = 8.6'$$
$$s = 15.4'$$
$$p = 56.4'$$

(Figure caption continued from page 118)

for new and full moons occurring near the solstices. Here, p is the moon's parallax. The alignments displayed here are those listed in table of 42 alignments analyzed in this study. Also, these ideal values are based on the assumption that the total sum, c_s, of curvature corrections is zero. This correction is a refinement that compensates for the fact that the real moon probably will not reach the greatest amplitude of the perturbation cycle at exactly the same time the 18.6-year cycle reaches its largest excursion. It also takes into account the unlikelihood that the observer will catch the moon on the horizon at exactly the time of maximum perturbation. The largest mean correction due to curvature is 0.6', and here the width of the arrows suggests the uncertainty generated by the curvature effects at various seasons of observation.

In summary, this diagram compares

$$\beta = \delta_o - \delta_e = \delta_o - (\varepsilon \pm i)$$

with

$$Q = \delta_e - (\varepsilon \pm i)$$

If there is good agreement between the two, the alignments (β)--represented again as gaussian areas--should cluster at significant values of Q along the horizontal axis. This clustering is observed, and it occurs at positions of observational interest in the perturbation cycle. We interpret this to mean that prehistoric moon watchers monitored even this subtle effect averaged over three or four 18.6-year cycles.

The most preferred declinations would have been those for maximum visible perturbation at major and minor standstills. These are framed by heavy lines in the diagram.

Using Broadbent's method, we evaluate the root mean square of the differences, $|\beta - Q|$, and determine that the chance of this distribution of β being random is one in ten thousand, or very low. (after A. Thom and A.S. Thom, Griffith Observatory) actually used, by looking at several examples in detail. One helpful site, Ballinaby, on the island of Islay, in Scotland's Inner Hebrides, consists of a single tall, slender slab of stone, about 17 feet high. Its base measures 3 feet by 9 inches, roughly, and its long dimension is oriented on the col 1¼ miles away. The col's declination agrees with the moon's upper limb at major standstill northern moonset ($\varepsilon + i + s$). Because the measured declination includes no perturbation component, we conclude the stone was used to mark the backsight for a time in the 18.6-year cycle when the major standstill season occurred more or less midway between the times of maximum and minimum perturbation. The period from one maximum to the next is 173 days. Minima occur midway between maxima, also every 173 days. Accordingly, times of zero perturbation take place every 86½ days.

Our study of the moon's motion tells us that at major standstill northern moonset we'll see the maximum perturbation at the equinoxes, with the moon in first or last quarter phase. Minimum perturbation coincides with the solstices and new or full moon. Ballinaby, by indicating a zero perturbation declination, must therefore be the average position between two extremes, one observed in June and the other in September. An average between September and December is also possible. The June and December moons would have been full and readily visible at the horizon. The September moon would have been a waning, third quarter moon. Its setting took place in the daytime, and the bright sky could have made the observation difficult.

Once the position of the Ballinaby stone was established by its erectors, it was used, we believe, to observe the setting, first quarter moon in March, near the vernal equinox, and the setting new moon (or, really, first crescent) in June, near the summer solstice. The former occurs at about 4:00 a.m., the latter about 10:00 p.m. In both instances, the sky is dark. The air temperature can be estimated: about 40° F. and 50° F. respectively. A sample calculation has already been included to show how we determine the exact declination of one of these lines.

There is ample room for side movement at

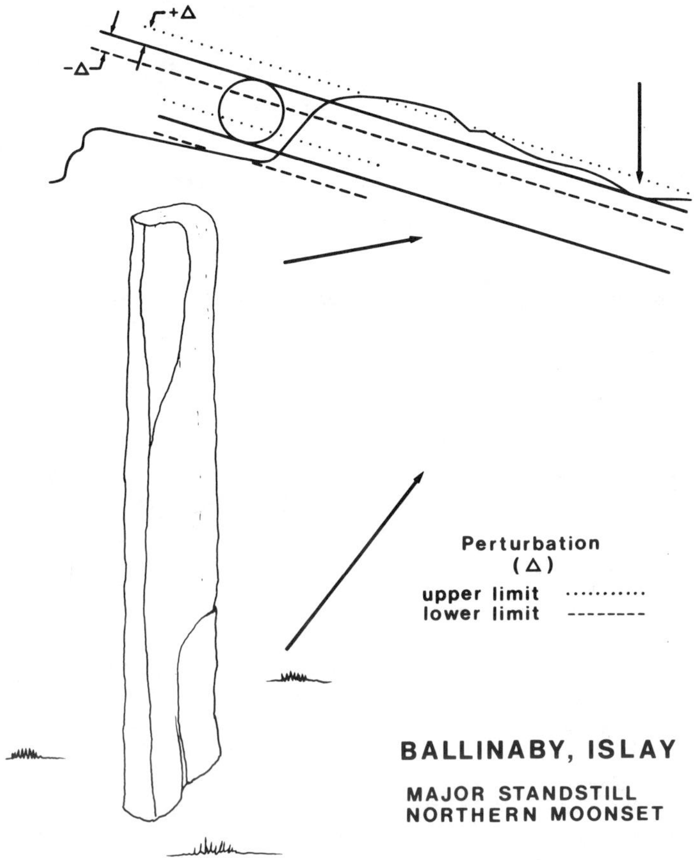

A tall, upright slab at Ballinaby, on the island of Islay, appears to indicate a V in the horizon profile of a peak $1\frac{1}{4}$ miles away. This line could have been used to observe the major standstill northern moonset at maximum perturbation (dotted line). After disappearing behind the peak, this moon would extend a little above the horizon again before disappearing altogether. Minimum perturbation on the position of the lower limb is also shown (broken line), as is the unperturbed moon (solid lines). We consider that the moon was observed in the lower notch, not from the 17-feet high menhir, but from a point to the southwest of it at vernal equinox and from a position to the northeast of it at summer solstice. The slab shows the middle position and hence represents the declination $(\varepsilon+i+s)$, with no Δ. (after A. Thom, Griffith Observatory)

HOW THE FORESIGHT WAS USED
BALLINABY, ISLAY

PATH OF SETTING MOON
MAJOR STANDSTILL NORTHERN MOONSET (FIRST QUARTER)

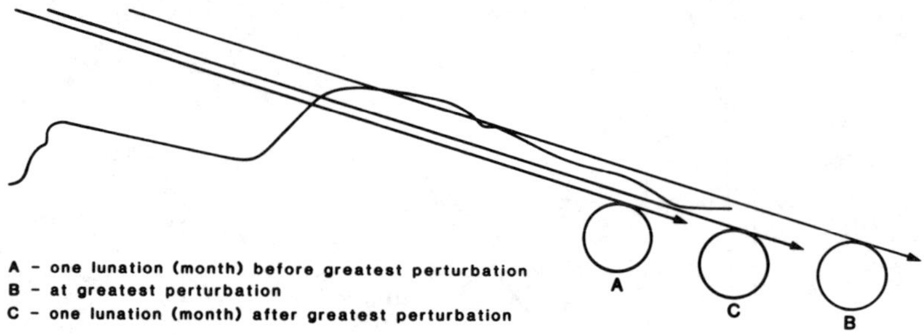

The position of the moon changes noticeably on successive lunations at the time of a standstill because the effect of perturbation shifts the moon relative to the horizon. Here, for Ballinaby, we see the path of the upper limb of the setting moon a month before maximum perturbation, at the perturbation peak--when actually observed from the menhir, and a month after maximum perturbation. Only at maximum perturbation is the upper limb far enough north to reappear on the descending north slope of the peak. (after A.S. Thom, Griffith Observatory)

Ballinaby, and this makes the site considerably more useful, and convincing, as a lunar observatory The position of the stone and the indicated foresight give a declination at the zero-perturbation point in the 173-day perturbation cycle. By standing some distance to the left of the stone, the notch would align with the maximum perturbation declination ($\varepsilon + i + \Delta + s$). That is, the upper limb of the major standstill northern moon, with maximum perturbation, would set behind Sliabh Mor, the crag to the northwest, and reappear briefly, a few moments later in the col.

Similarly, a position some distance to the right of the tall Ballinaby menhir, aligns with the same moon at minimum perturbation (declination = $\varepsilon + i - \Delta + s$).

One could, for example, stand at what would have been a known spot left of the stone and observe three successive lunations at the time of major standstill. The northernmost moonset a month before the March or September extreme falls below the col. The moon passes behind the hill, but it is not seen again. The next month, the northernmost moon sets behind the hill and *does* reappear for a moment farther down the slope. A month later the moon repeats the performance of the first of the three lunations. The lunation with maximum perturbation, could, in this way, be determined.

By standing at the stone itself, 87 days or so after the time of maximum perturbation,observers would see the zero-perturbation moonset in the col. A month later, the northern moonset is below it.

This procedure could be repeated over many months at the time of major standstill, and it would provide ample information on the moon's progress through the 173-day perturbation cycle. Another irregularity, higher up the slope on the foresight, also could have been used to observe the moon at maximum perturbation from the Ballinaby stone. The zero-perturbation line is the most convincing, however, and it suggests an interest in the midpoints in the perturbation cycle. There are other sites at which observations of both the perturbation maximum and midpoint are possible (Campbeltown and Temple Wood). If bad weather prevented

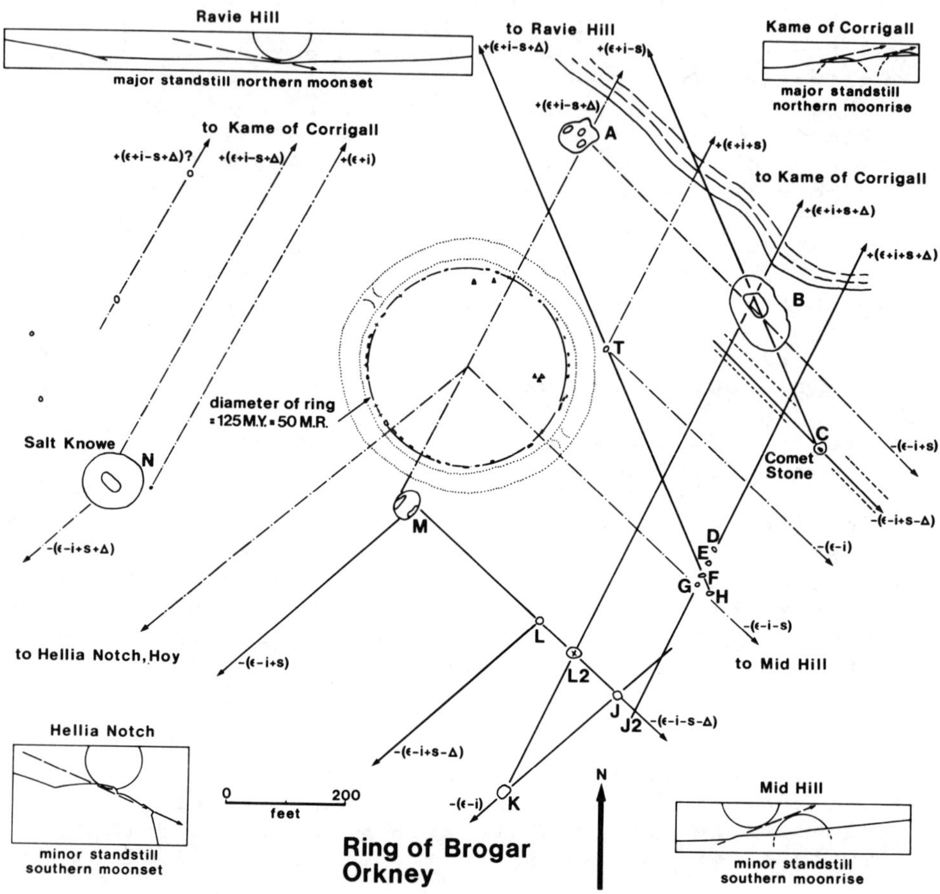

In Orkney, at the Ring of Brogar, several mounds may combine with horizon features to mark four lunar standstill alignments. (after A. Thom and A.S. Thom, Griffith Observatory)

determining the lunation of maximum perturbation, the cycle might still be established and verified through observations of other significant declinations (midpoint and minimum perturbation).

Now that we have considered Ballinaby in some detail, we can see how some of the other sites may have been used. At the Ring of Brogar, on Mainland, Orkney, two lines to the northwest--and irregularities on the slope of Kame of Corrigall, about 5 miles away--are indicated by alignments between earthen mounds situated outside the stone circle and its henge. These lines agree with the declination of the northernmost rising moon at major standstill *and* at maximum perturbation (declination = $\delta = \varepsilon + i + \Delta + s$). One lunation before the perturbation maximum, the waning third quarter moon rises in the northeast in a dark, nightime sky. The line from mound L_2 and across mound B continues to the horizon, but the rising moon first appears to the right of the chosen foresight. One month passes, and now, close to the autumnal equinox, the northernmost moon reveals its first gleam above the small bump on the horizon. One more month passes, and the moon again rises to the right and east, of the foresight. The second line to the Kame of Corrigall is formed by mounds G, F, E, and D, and gives an almost identical declination.

From Brogar the cliffs at Hellia, on the island of Hoy, permit precise observation of the minor standstill southern moonset. There are two notches on the upper slope of the cliffs, and both provide the appropriate declination for lower limb of the full moon. The foresight is 8.3 miles southwest of the ring and its mounds, which may indicate three possible declinations in the perturbation cycle. Mound J and mound K align with Hellia for the perturbation midpoint and the center of the moon's disk ($-(\varepsilon-i)$). Standing at mound L, the line to Hellia agrees with either of two possibilities:

a. the center of the moon's disk at maximum perturbation ($-(\varepsilon-i+\Delta)$)
   December full moon, near winter solstice

b. lower limb of the moon at minimum perturbation ($-(\varepsilon-i-\Delta+s)$)

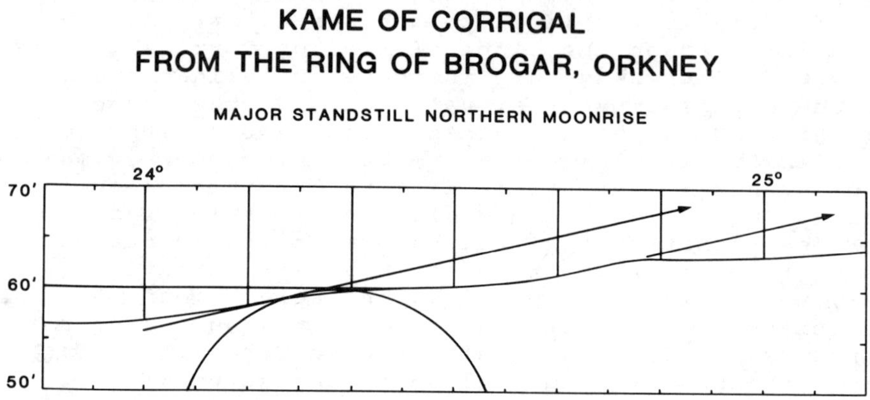

Alignments from Brogar to the Kame of Corrigall foresight could have targeted major standstill northern moonrise at maximum perturbation. The arrow on the left corresponds to the view from mound K and over mounds L2 and B. The arrow on the right corresponds to the view from a point beside J and on the extension of the line formed by mounds D, E, F, and G. Both give the same maximum perturbation declination. (after A. Thom and A.S. Thom, Griffith Observatory)

HOW THE FORESIGHT WAS USED
KAME OF CORRIGAL
AS SEEN FROM THE RING OF BROGAR, ORKNEY

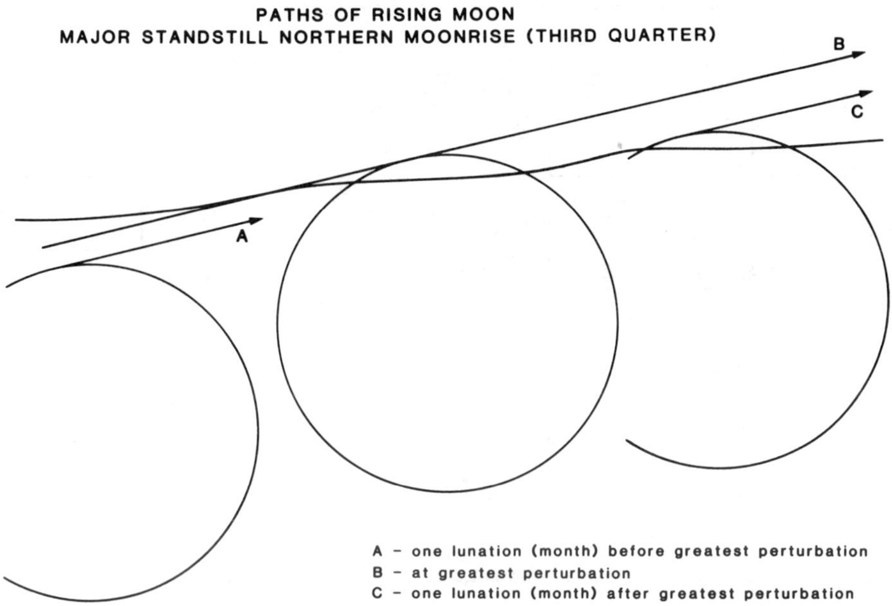

The monthly change in the position and path of major standstill northern moonrise can be detected with respect to the northeast horizon features of Kame of Corrigall. The view here shows the circumstances one month before the perturbation maximum, on the perturbation maximum, and one month after perturbation maximum, as seen from mound L_2 and over mound B. (after A.S. Thom, Griffith Observatory)

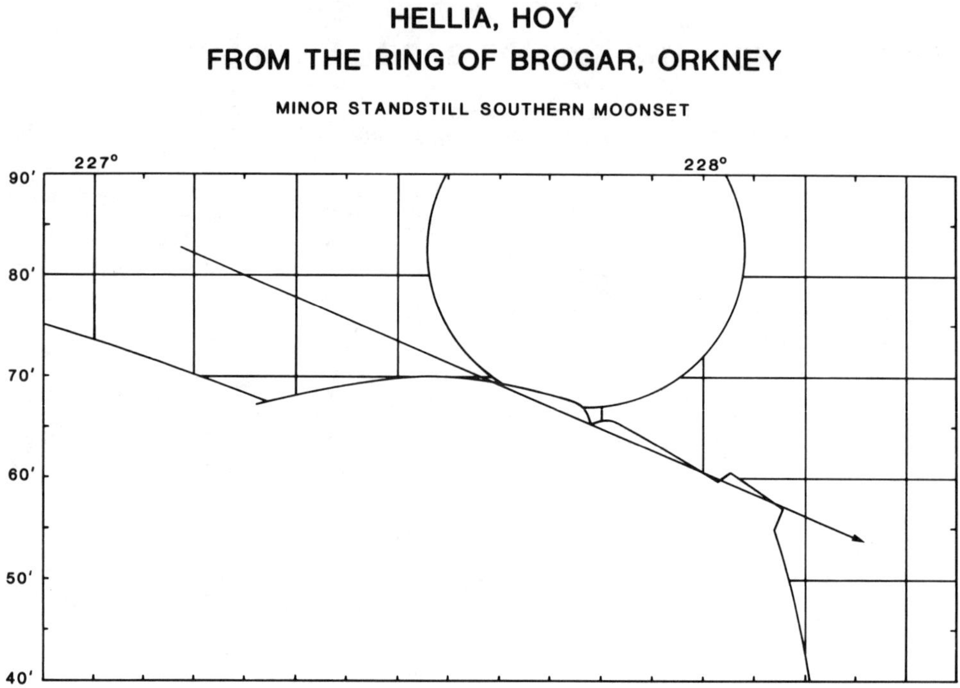

Minor standstill southern moonset would be monitored over the cliffs of Hellia on the island of Hoy by an observer at the Ring of Brogar. From different backsights, marked by mounds, different stations in the perturbation cycle could have been detected (after A. Thom and A.S. Thom, Griffith Observatory)

HOW THE FORESIGHT WAS USED
HELLIA, HOY
AS SEEN FROM THE RING OF BROGAR, ORKNEY

PATHS OF SETTING MOON
MINOR STANDSTILL SOUTHERN MOONSET (FULL)

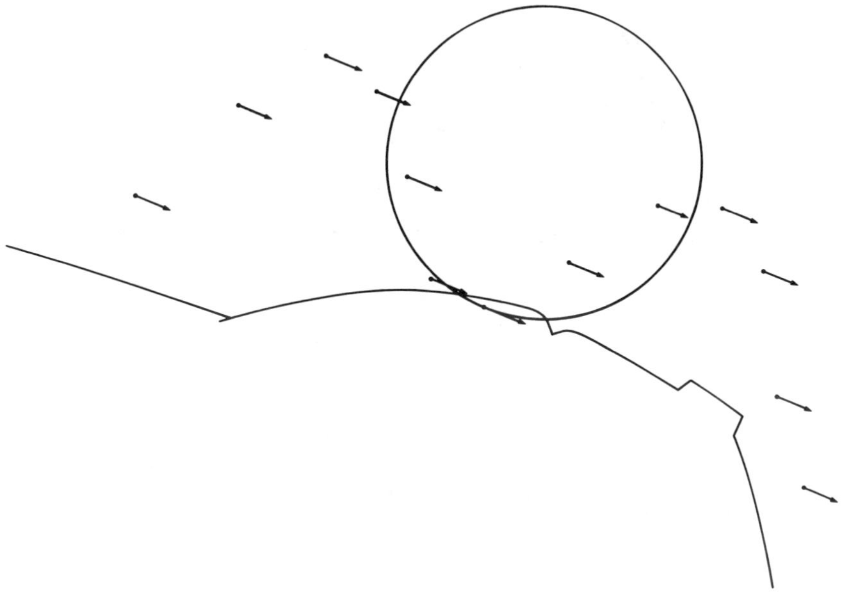

We must imagine that from some point near the stone circle, prehistoric moon watchers would have been able to catch the December full moon's lower limb just slipping along the Hellia slope. This would occur at maximum perturbation, with $\delta=-(\varepsilon-i+\Delta+s)$. This path and 11 other arrows shown here represent the paths of the monthly southern extremes of 12 successive lunations framing the minor standstill. They demonstrate that only December's full moon will "roll" down the cliff. By watching the moon carefully, the changes in distance between its monthly southern limit and the Hellia foresight, could be detected. This, in turn, could establish the date of perturbation maximum and settle the placement of the various backsights. (after A.S. Thom, Griffith Observatory)

RING OF STENNESS, ORKNEY

MINOR STANDSTILL NORTHERN MOONSET

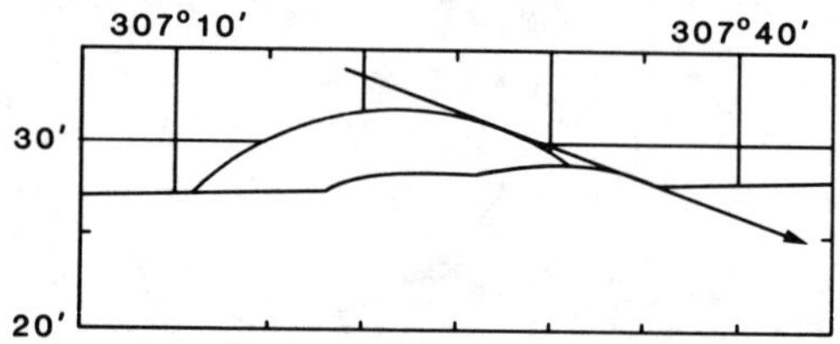

A cairn northwest of the center of the Ring of Stenness, Orkney, is the foresight for the minor standstill northern moonset. The moon's upper limb just reaches the northern tip of the cairn at maximum perturbation. (after A. Thom and A.S. Thom, Griffith Observatory)

quarter moon, near equinox (probably waxing, first quarter moon, in September).

A third line, from ground level beside mound M, agrees with the lower limb of the setting moon at the zero-perturbation point $(-(\varepsilon-i+s))$.

Just as at Ballinaby, we find sufficient open and relatively flat ground at Brogar to have allowed the observers to alter position for various times in the perturbation cycle. There is also enough room to have permitted simultaneous observations of both upper and lower limbs on the same foresight. Averaging between these positions could have given the locations that indicate the center of the disk only where there is ample room to observe on several different lines.

At minor standstill then, the prehistoric moonwatchers of Brogar might have looked to Hellia notch on Hoy for the extreme point of the setting moon each month. We imagine them to have been on the lookout for the innermost extreme, the third quarter moon near the vernal equinox or the first quarter moon near the autumnal equinox. These moons slice down through the sky well above the Hellia profile, however, and the megalithic astronomers would have had to catch the minimum perturbation, visible in the position of the December full moon, which would have slid right along the top of Hellia, its lower limb just grazing through the notches in the profile.

We infer from the alignments we believe to be preserved that the prehistoric observers knew about the cycle, at least insofar as it affected the position of the setting moon at the time of the standstills. They would have watched each lunation's southwestern extreme and would have been perfectly aware of the variations.

Near the Ring of Brogar is what is probably an older stone circle, the Ring of Stenness. A cairn, some distance to the northwest, beyond the Ring of Brogar and on the horizon, provides another lunar line in concert with the center point of the Stenness circle. From Stenness, the upper limb of the minor standstill northern moon sets behind the cairn, its last gleam just touching the cairn's northernmost edge. The exact declination agrees with the maximum perturbation $(\delta = \varepsilon - i + \Delta + s)$

HOW THE FORESIGHT WAS USED FROM THE RING OF STENNESS, ORKNEY

PATHS OF SETTING MOON
MINOR STANDSTILL NORTHERN MOONSET (FULL)

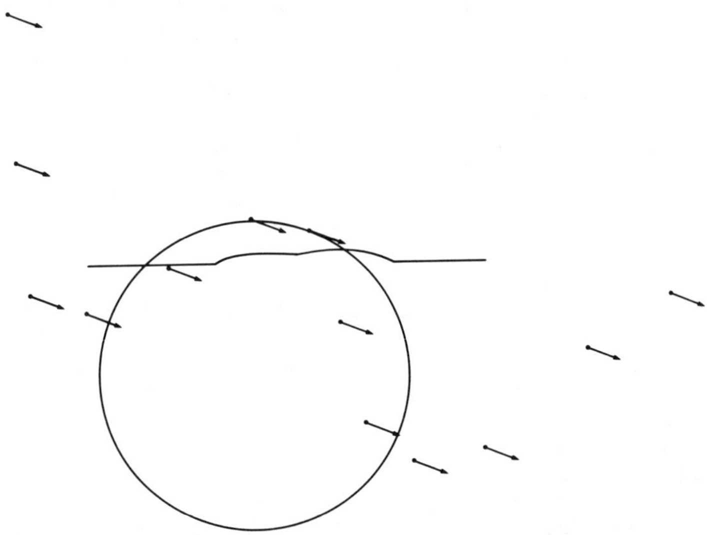

The effect of perturbation on the position of the minor standstill northern moonset at Stenness can be seen in the 13 arrows that mark the paths of the upper limb of the setting moon in the six months preceding and the six months following the minor standstill. The upper limb of the full moon nearest the summer solstice contacts the edge of the cairn at the monthly extreme that coincides with the perturbation maximum. At this time, the declination of the upper limb is $\delta=(\varepsilon-i+\Delta+s)$. Here, the perturbation amplitude is scaled to the horizon and moon. The arrows for the monthly extremes have been spaced to mimic the wave of perturbation itself. (after A.S. Thom, Griffith Observatory)

and generates a value of ε, the obliquity of the ecliptic, identical to that obtained from the northern declinations in the Brogar alignments. We believe the absence of a natural foresight for this alignment at Brogar, as well as an environment too awkward there for the construction of artificial foresight, prompted observation from Stenness.

The upper limb of the setting moon approaches closer and closer to the right hand edge of the cairn until December, when the full moon--at maximum perturbation, just grazes it. For the next several months the moonset backs away from that position as the shift back toward minimum perturbation (and lower declination) shows up.

Determination of the eclipse seasons from the date of the innermost moonset actually makes the Stenness line superfluous in the neighborhood of Brogar. There is already enough here to keep track of the day count and the moon's position in it. We can only infer that the observers were putting on record a variety of positions involved with various times in the perturbation cycle. We wonder, therefore, whether many other "redundant" sites are yet to be discovered.

Finding the Perturbation

We have considered several individual sites to show how the position of the moon on the horizon, could, at either standstill, permit megalithic skywatchers to monitor the variations introduced by the perturbation cycle. In our calculations we have evaluated the perturbation, Δ, according to the equation

$$\Delta = 519'' \cos 2(\lambda' - \Omega) - 42'' \cos 2(\lambda - \lambda') + 39'' \cos 2(\lambda - \Omega),$$

where λ = longitude of the moon,

λ' = longitude of the sun,

Ω = longitude of the node.

(This formula was provided by L.V. Morrison of the Royal Greenwich Observatory--private communication) and the resulting values of Δ differ accordingly from those in *Megalithic Remains in Britain and Britanny*.

Celestial longitude is measured counterclockwise (eastward) along the ecliptic from the First Point of Aries (the vernal equinox). At major standstill, when the moon is at an extreme declination, the longitude of the moon is 90° when north and 270° when south. The node has a longitude of 0° or 180°, respectively. Only when the node is in conjunction with or in opposition to the sun can the moon's maximum declination attain its greatest possible values. These times are separated by 173.3 days, or half of the eclipse year. When the node is near the First Point of Aries, the sun's longitude is near 0° or 180°. Therefore, the moon at greatest declination extremes requires

a. the moon's longitude be 90° or 270° and

b. the sun's longitude be near 0° or 180°.

This means the moon is at first or third quarter at its declination extreme. Also this occurs near the vernal or autumnal equinox. Probably we should choose the solar longitude that places the lunar event in a dark sky.

When we examine the moon's monthly declination extremes at the major standstill of 1969, we see that the perturbation maxima occurred at the autumnal equinox, 1968, the vernal equinox, 1969, and the autumnal equinox, 1969. Minimum perturbation happened at the solstices, in December and June. Zero perturbation occurred twice as often, midway between every solstice and equinox.

The most important events, perturbation maxima, never take place at new or full moon. The disk at the quarter moon is not fully illuminated, but it is illuminated enough to permit the observation.

At the equinoxes, the first and third quarter moons at perturbation maximum experience a $\Delta = +8.6$ arcminutes. At the solstices, the new and full moons are shifted by $-\Delta = -10.0$ arcminutes, perturbation minimum. At perturbation minimum the moon's longitude is 90° or 270°, but the sun is also at 90° or 270°, and not at 0° or 180°, as at perturbation maximum.

We know of only three alignments which show directly lunar declinations at the solstices:

Callanish V, on the Isle of Lewis; Stenness, in the Orkneys; and Crois Mhic Aoida ("McKie's Cross"), on Kintyre. They appear to incorporate the effect of maximum perturbation, as sighted on the moon's upper limb. Although we feel these three sites should not be ignored, the probability is low that precise observations were consummated at the solstices. We have not considered, therefore, declinations given by other combinations of the moon's semi-diameter, perturbation, and standstill cycle for the solstice seasons.

Every standstill is different in detail, but at each the moon approaches the same maximum declination. In general, the prehistoric astronomers would miss the absolute maximum. It probably occurred most often between the two observing times, 173 days apart.

All lunar calculations also require an estimate of the effect of parallax and a value for the radius (semi-diameter) of the moon. Because we do not know the exact date when each monument was built, we must assume a mean value of 56.4 arcminutes for parallax at the equinoxes and 57.4 arcminutes for the solstices. Similarly a mean value of 15.4 arcminutes was adopted for the moon's semi-diameter at the equinoxes and 15.6 arcminutes at the solstices. L.V. Morrison advised us of these values.

How to Find the Lunar Extreme if it Doesn't Always Occur at Moonrise or Moonset

The moon can only be observed rising, or setting, once each day. Between consecutive moonrises, occurring perhaps before and after the time of maximum declination, the moon's declination can change quickly and considerably. The observer will miss, therefore, most of the actual declination extremes. We believe this problem could have been overcome by extrapolating the extreme position from the points actually observed. The procedure has been fully described in *Megalithic Lunar Observatories* and in "The Uses of the Alignments at Le Menec Carnac" (*Journal for the History of Astronomy, Vol. 3*, Part 3, No. 8, October, 1972, pp. 151-164), and so only an outline will be given here.

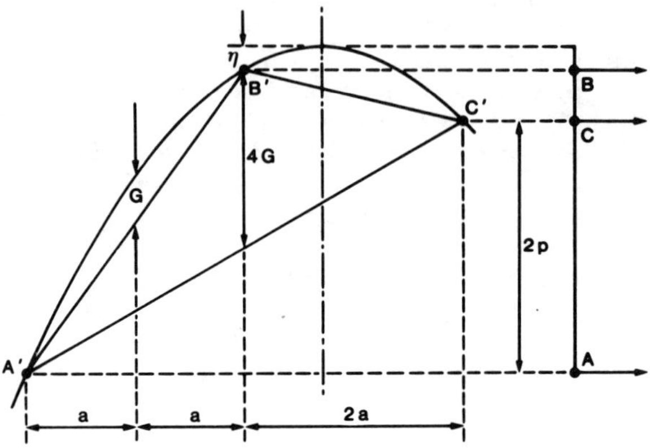

3 stakes $\eta = p^2/16G$

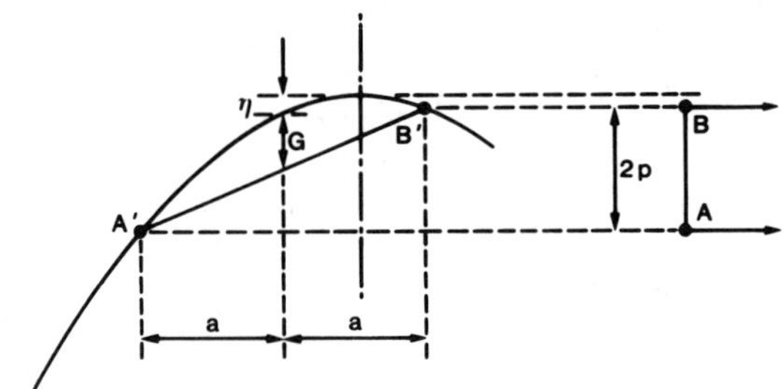

2 stakes $\eta = p^2/4G$

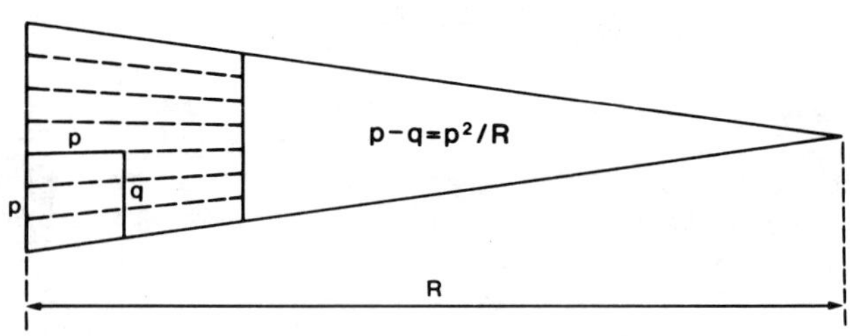

Because the moon may not have reached its greatest excursion at exactly the time it is seen on the horizon, megalithic astronomers would have had to devise a way to ascertain the correct place to stand for the true extreme from the places from which the moon was actually observed to set. One method requires three successive observations that place the moon behind the same foresight. Three stakes mark the backsights (A, B, and C). Points on a parabolic curve can be shown to correspond with these positions, and the geometry of the situation is shown in the drawing at the top. It can be shown that the displacement beyond B for the true extreme is given by

$$\eta = p^2/16G,$$

where p is half the distance between the first and third stakes and where G is the distance on the ground that a half-day's movement from declination maximum would require.

A simpler technique involves two observations, and the situation is shown in the drawing below, where A and B mark the two observation backsights. In this circumstance, the additional displacement from B to the true extreme's backsight is

$$\eta = p^2/4G.$$

Fans of stones, laid out on the ground in a configuration like that in the lowest drawing and to the correct dimensions (determined empirically, by trial-and-error), could then be used in subsequent lunar observations to help obtain the correct position for an ongoing program of extrapolating the true position for the monthly extreme, month after month. (after A. Thom and A.S. Thom, Griffith Observatory)

We are considering the decrease from maximum in lunar declination that may occur in half a lunar day. The lunar day is close to the interval of time between two successive moonsets (or moonrises). From one moonset to the next an observer must shift his or her position on the ground by some amount to keep the moonset aligned with some distant horizon feature. The size of this shift depends upon the distance to the horizon and the change in azimuth produced by some unit change in declination.

On three successive nights an observer will mark the spots from which the moonset was seen to align with foresight. Because the decrease in declination increases with the square of the interval of time elapsed from it, we can establish a detailed relationship between positions of the three stakes that mark the three spots at which moonrise was observed. We are looking for the extra distance we would have had to move to the left, say, beyond our leftmost stake to see the actual lunar extreme, had it occurred on the horizon. In fact this can be evaluated, for it can be shown geometrically and analytically to be related to the distance between the first and third stakes and the distance on the ground a half-day's movement from declination maximum would require. Somehow, this last quantity had to be determined empirically. Once known, the correction to the observed stake positions could be obtained from sectors or triangles laid out in stones to the proper dimensions and design.

Geometric analysis tells us that it would be sensible to require the radius of curvature of a sector to be 8 times the distance an observer must move, on level ground, to preserve the moon's alignment with a distant horizon feature, from one moonset, say, to the next. Of course, this displacement is related to the distance to the horizon in a specific way, but field work and practical observation would provide the required information. Algebraic analysis is not a prerequisite for application of the technique.

Fans for the Moon

It is easier to visualize the process by considering the case in which only two observations are made, at two moonsets occurring before and after the month's maximum declination. In this case the radius of curvature of the grid must be 2 times a day's displacement on the ground.

First we measure out half the displacement, from the corner of a sector, laid out on the ground in stones perhaps, just for this purpose. We measure out this length along the sector's base, across its radii. Then we lay out the same distance along a radius, toward the sector's center of curvature. Finally we measure back across the radii to the bounding radius of the grid. Because the sector is tapered, this last distance is less than that set out at the grid's base. The difference between these two lengths can be obtained by comparing them, physically, and this difference is the length we seek. It is the amount we must add to the position of the stake nearest the spot with which the moon at maximum declination would have aligned, had maximum occurred at moonset. Once this spot is found, it can be marked permanently. An analogous procedure can be applied to observations of three successive moonsets for a more refined result.

Fans of stones set out in the proper dimension for use as extrapolation grids, as described above, do exist. There are four in Caithness, Scotland, and two near Carnac, in Brittany. All of them have radii that suit, at least approximately, the distances to the sites' proposed lunar foresights. None of these seems to have, however, the large radius appropriate for application of the "three moonset" technique, and we must suppose that approach was not normally taken.

A large megalithic observatory, as we imagine was centered on Le Grand Menhir Brisé at Locmariaquer in Brittany, would have required, in turn, very large displacement shifts on the ground, perhaps too large. It is possible this so worried the builders they attempted to develop an alternate method, one which, in fact, makes use of the alignments at Le Ménec as we find them. This technique could have dealt with variations in the displacement and is described in our paper on the uses of Le Menec alignments. The alignments could have served most of the proposed backsights, but we have no way of showing that they were so used.

Existing sectors in Caithness are all built of small stones, so small that on land destined for agriculture they could have been cleared away without leaving a trace. Temple Wood, near Kilmartin, in Argyll, Scotland, and Brogar, in Orkney, have such promising lunar alignments, it may be that auxiliary stone grids were removed from their neighborhoods. It is possible, too, that a simpler extrapolation technique was used--one that involves a triangle with a base equal to 2 times the shift of an observer's position, from one moonset to the next. Certainly some extrapolation method must have been available. Otherwise the lunar observatory would have been useless as a tool for precise work.

Problems and Possibilities

Even though the moon would rarely reach a declination extreme at the exact moment of its rising or setting, an empirical and relatively simple system of extrapolation, based upon two, or perhaps three, successive appearances of the moon on the horizon, could have allowed the correct backsight for, and date of, the declination extreme to be found. Critics have argued, however, that clouds and bad weather might often have interfered with observation and so doomed the entire process. This need not be so. If a day's moonsighting be lost, an empirical approximation, determined through previous experience, would allow a reasonably close assessment of the position an observer should have occupied on that day. For any site, there is a simple relationship between the positions of stakes placed, say, for three successive moonsets bracketing the time of each month's declination extreme. The distance from the stake for the second moonrise to the midpoint between the stakes for first and third is the same, from month to month. The exact positions of the stakes shift each month as the moon's declination gradually changes, but this relationship between the three stakes is identical. Although on the ground distance to be shifted from the midpoint of first and third moonsets varies from site to site, at all sites it is the shift equivalent to 48 arcminutes displacement on the horizon.

Similarly, if all observations during the month of maximum declination be lost, the correct position of the backsight can still be surmised from the extreme stake positions obtained at the declination extrema for the previous and following lunations. Observers accustomed to following the moon's behavior over long periods could have discovered this alternative with relative ease.

We should keep in mind that there are other basic obstacles to acquisition of the information we propose the megalithic "astronomers" were seeking. For example, it is much easier to observe the setting moon than the rising moon. Near moonset, the moon can be followed along the horizon until it vanishes. By contrast, the rising moon appears suddenly. We propose, however, that moonrises, as well as moonsets, were used to ascertain the backsights for the declination extrema. How could moonrises have been mastered?

A second observer standing on flat ground nearby the primary moonwatcher, can be positioned to spot the moon coming up over a sloping horizon--and so warn the chief observer to be ready. If the ground rises behind the chief astronomer, the helper need only be above the backsight position, on the hillside, to shout a warning that the moon is coming up. This we believe was the purpose of the stone on the hillside above the "observing platform" at Kintraw. Although there the winter solstice sunset was the target, the same technique would have been useful.

In a similar way, the sun's green flash could have been a criterion sufficient to assure the observer that the upper limb--and therefore the exact moment of rising or setting--had been sighted.

An early warning of the imminent moonrise would have permitted the observers to anticipate and so pinpoint the location of the first gleam of the moon's upper limb. Otherwise they would have had to wait until a substantial portion of the moon had appeared.

At Brogar, mounds (Salt Knowe and Mound A) were erected at positions which turn out to be special. From these, warning could be given that the moon was about to emerge from behind the Kame of Corrigall on the one hand and Mid Hill on the other. Similarly the special row of boulders on the hill behind Mid Clyth could have served the same purpose.

At the time when the moon was expected to rise above Mid Hill, which slopes up towards the south, a helper was needed to give the warning. Ideally, the assistant would have been stationed northeast of the appropriate backsights, but the shores of Loch Harray limit the distance to which the helper could be removed. To compensate, the eye-level of the assistant could be raised, and this may explain why Mound A is as far to the northeast as it could be built and as large and as high as it is. Likely it was desirable, in partial darkness, to provide some indicator of the direction in which the small notch on Mid Hill lay. Posts on Mounds A and B, coincident with the line-of-sight to the Mid Hill foresight, could have solved this problem.

For the Kame of Corrigall moonrise, Salt Knowe provided a spot for early warning. The mound is placed well to the left of the appropriate backsights but not so far left that it occupies the lower ground to the west. The presence of these mounds at Brogar greatly strengthens the case that we are dealing with a site at which practical observation took place. Be that as it may, we are convinced the setting moon was preferred. In our 42 best alignments there are 28 for moonsets and only 14 for moonrises.

Yet another argument supports the idea that genuine solar and lunar observatories existed. At some of these sites we find evidence of the use of several foresights, all seen from closely placed backsights. Multiple moonlines occur, for example, at Stonehenge, Temple Wood, and Brogar. Ballochroy has two possible alignments on the sun. At Le Ménec, stone S is the backsight for both a solar line and a lunar line. Another situation also occurs; some potential observatories involve several backsights and a universal foresight. The summit of Boreray, the most northerly island of the St. Kilda archipelago in the North Atlantic, is a foresight for the setting sun from four backsights: on South Uist, Benbecula, North Uist, and Harris. All of these alignments coincide with primary calendar declinations. In the Carnac area, both the Manio stone and the Grand Menhir Brisé act as foresights for numerous backsights.

Megalithic Observatories: When and Why

Each lunar alignment provides an opportunity to assess the value of the obliquity of the ecliptic, ε. Because this angle slowly changes, the value calculated implies a date of construction and use. We have corrected our lunar declinations for the assumed values of perturbation, semi-diameter, and orbital inclination appropriate to each moon line. The average obliquity that results corresponds to a date of 1575 ± 100 B.C. No greater precision is possible because parallax varies and so alters the calculated declination of the moon. Mean parallax has been used. Also, it is difficult to estimate the effect of refraction at low altitudes. The graze effect itself varies from a mean. Furthermore, faulty measurements of altitude and azimuth made by ourselves and errors made by the prehistoric monument builders in the final choice of backsight positions are virtually inevitable. Geological movement, erosion, and vegetation growth all may complicate things further. All of these sources of uncertainty must contribute to the inaccuracy of the measured declinations. It is amazing therefore that the root mean square residual deviation of our 42 best lunar alignments from their expected declinations is as small, 1.34 arcminutes, as it is. This suggests that the final backsights were refined only after a long period of observation. We estimate that these correct, average positions could have been found after perhaps ten standstills, or about 84 years.

The moon's declination is the object of our calculations, and we obtain it from the azimuth indicated by the alignment. The system of right ascension and declination is centered on the center of the earth, but azimuth observations are made at the earth's surface, an earth radius away from the center of the celestial coordinate system. We can correct our calculated declinations by this small angle of horizontal parallax, but its actual amount depends on both the size of the earth and the distance to the moon. The moon's distance changes because the moon's orbit is elliptical.

We have discovered that although the parallax of the moon at the time of standstills is not constant, it varies in a peculiar, sinusoidal cycle. In 179 years the parallax will vary with an ampli-

CALCULATED DECLINATIONS FOR THE MOON AT STANDSTILLS 2100 – 1300 B.C. (WORMADALE HILL)

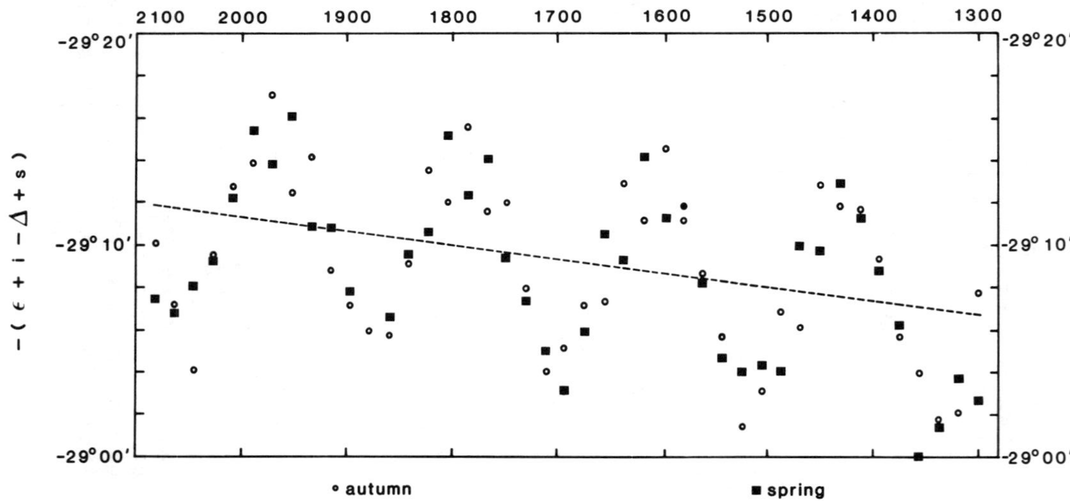

The moon's major standstill southern declination between 2100 B.C. and 1300 B.C. numerically decreases by a steady 39 arcseconds per century. This is shown here for Wormadale Hill, Shetland, and the moon's upper limb at minimum perturbation (dotted line: $\delta=-(\varepsilon+i-\Delta+s)$) On top of this change, a cyclical variation in the moon's parallax with a 179-year period is superimposed. These Greenwich calculations can be compared with the measured declination of the Wormadale Hill alignment for an assessment of the site's age. (after A. Thom and A.S. Thom, Griffith Observatory)

tude of 3 arcminutes above and below the mean line. Calculations by A.T. Sinclair of the Royal Greenwich Observatory confirm this. If we assume that megalithic lunar alignments reflect a scatter of observations through the 179-year period of parallax variation, we can extract an estimate of the root mean square variation due to parallax. It is 2.12 arcminutes. We can also calculate the root mean square variation due to parallax that would emerge from observations of lunar standstills at a given site (in this case, Wormadale Hill, in Shetland). For 20 standstills between 1800 and 1630 B.C., the variation is between 3.0 and 3.5 arcminutes. We recall that the root mean square deviation of our 42 lunar lines from their expected deviations is 1.34 arcminutes. What is interesting is that this is significantly less than the variation automatically introduced by the sinusoidal variation in parallax. In some way the megalithic moonwatchers must have compensated for this effect.

We propose that work at the megalithic lunar observatories extended over a long period of time, longer than, say, three or four standstills. Other wise the root mean square deviation we find in the alignments could not be so low. If, for example, the alignment at one site is the product of three or four standstills of observation centered perhaps on 1600 B.C. and another alignment derives from a comparable interval centered on 1500 B.C., a difference of 8 arcminutes in declination for the same configuration of the moon would occur. Our results show, however, that the differences in the actual alignments are far less. The scatter in the indicated declinations from those expected, based upon a mean estimate of parallax, is so small that we conclude existing alignments were not products of individual standstills. Instead, we guess the megalithic moonwatchers took the mean position, on the ground, of a number of observations made at several standstills throughout a period of at least 150 years. This implies a stabilized, structured means of obtaining, preserving, and communicating this rather specialized information. If we knew, by some other means, the dates at which various centers were operative, matters might be much clearer.

For example, we cannot determine the date of Brogar through a precise solar alignment, and the obliquity of the ecliptic it would imply, because so far no reliable solar line has been found near to the site. Local people have pointed out to us two or three places where they had either seen a standing stone years ago or had been told of its previous existence, and there are other standing stones that survive in the district. None of these possibilities sheds light on the site's date, however.

It is of interest here to note that the uncalibrated radiocarbon date for Newgrange in Ireland averages about 2470 b.c., which gives a date of about 3000 B.C., and it is on record that rays from the winter solstice rising sun shone through a suitably oriented shaft into the central chamber.

The obstacles to accurate dating on the basis of lunar alignments have already been enumerated. The best date we can obtain for the use of Brogar as a lunar observatory is 1485 ± 100 B.C.

Archaeologist Colin Renfrew has excavated the ditch that surrounds the platform on which the stones of the Brogar circle stand, but no datable material has been found. Organic material in the bottom of the ditch cut into the rock around the Ring of Stenness was found by Dr. Graham Ritchie, and with it he dated this nearby sister monument of the Ring of Brogar to 2356 ± 65 b.c. Charcoal from the construction in the center of the ring provided a date of 2238 ± 70 b.c. When recalibrated, these radiocarbon dates average to about 2960 B.C., early in the third millennium B.C. Brogar, by its proximity, is not likely to be much younger.

We may surmise that in the beginning the Brogar site was chosen to occupy a spot from which three or four foresights for standstill declinations of the moon may be seen. At this early stage however, the fine accuracy we see at the site was not developed. As experience and knowledge were accumulated we hypothesize that the other features of the Brogar observatory, the mounds and stones that permit fine tuning to the moon's behavior, were added. The precise lunar alignments we find at Brogar today may be more than a thousand years more recent than the ring itself. Because the Brogar ring may be of the same date as Newgrange (about 3000 B.C.) with its winter solstice-aligned passage, the above surmise may not be far from wrong.

Over the years we have surveyed more and more sites. The subject of megalithic lunar observatories has evolved, and our technique of data presentation has developed and expanded. For this reason, data reported here differ from previously published accounts.

In 1967 we felt we had 40 lunar lines worthy of consideration. At that time declinations were determined to one-tenth of a degree, and alignments which deviated from the expected declination by more than one degree were excluded. By 1971, many of the sites had been resurveyed, and many new alignments were found. A total of 80 lines had, by then, been investigated. Of the original 40, 28 were discarded, and 52 remained with declinations established to one-tenth arcminute.

Since 1971, we have added 15 more lunar lines: nine at Brogar, two at Ballinaby, and one each at Stenness, Dunskeig, Skipness, and Wormadale Hill. The total number of lines examined to date is now 95. Here we retain 42 of them in our analysis and justify inclusion or rejection throughout. In general, we omit all sites for which the backsight's position is not known with sufficient accuracy.

Of those alignments retained, we have been criticized for including Blakely Moss, in Cumbria. This ring was reconstructed in 1925, when eight of the original thirteen stones were missing and the remaining five were reerected. We cannot be absolutely certain that there was a prehistoric ring there. With five stones lying in a curve, however, a reerector would have had a good idea of the required diameter. It is likely the base of each stone as well as each stone's hole was found before putting the stone upright. The distance to Screel Hill, the foresight, is 30.7 miles. A shift in position equal to half of the circle's diameter alters the azimuth of Screel Hill by 0.57 arcminutes and the observed declination by about 0.17 arcminutes. This is still sufficiently accurate. Screel Hill was seen from the ring, but the sun did not appear when we visited the site. Altitude and azimuth were taken, therefore, from the one-inch Ordnance Survey map.

In any case, we include here tables which list the 42 lines on which our analysis is based and the 25 lines deemed unacceptable since the 1971 study. Of the 95 alignments examined, a total of 53 have been judged inadequate for the purposes of this analysis. (See Appendix Tables 1-3.)

Suppositions About the Past

We must, we think, avoid overly restrictive and categorized judgments about what kind of foresights and backsights prehistoric astronomers may have used. We are four thousand years too late to know with certainty, and we shall never have a fraction of the experience they had in selecting and using the lines they did. We must learn by examining each and every potential example we can find, and we now have enough information about enough sites to allow us to imagine how the sites were used.

We know that there is no clear connection between the actual date of the standstill and the date of the observed maximum declination. An extrapolation method was absolutely necessary, but this, we argue, was in the capacity of the megalith builders of the mid-second millennium. Observations were made, we feel, with care, and the information was, we judge, in some way preserved. The lunar observatories must have been refined over a long period of time.

Until recently it was unthinkable that our rude forefathers, living 120 generations ago in small, isolated communities--illiterate peasant farmers and hunters--could have possessed the intellect and skills demanded by the observational and geometrical techniques attributed to them. But we have collected evidence from Brittany to Shetland, from Stonehenge to Callanish, to support our interpretation of neolithic and bronze age "science." Any evaluation of the megalith builders' achievements in astronomy, geometry, and engineering must reconcile this evidence.

We believe that the people of northwest Europe four thousand years ago had a basic knowledge of applied mathematics and astronomy. The intelligence required is of the same order as that needed to develop a first class language, with vocabulary, grammar, and syntax. Classicists maintain that the most perfect language of all probably is ancient Greek, which obeyed a complex set of rules. Greek existed three thousand years ago. Where did it originate? How long did it take to develop? We cannot tell. We infer, however, that prehistoric people in northwest Europe were able to develop in the same way a working knowledge of elementary astronomy. Why should their intellectual abilities have been any less than those of the early Greeks, or their predecessors, who left the Greek language?

Description of megalithic lunar astronomy is necessarily extremely complicated because the moon's motion is so intricate. We now know the precise behavior of the moon. In prehistoric times, the moon's movements could be monitored only by observing on the horizon and only at each maximum or minimum turning point of its cycle of change of azimuth. It is extremely unlikely that the megalithic moonwatchers ever used the concept of declination. They could only observe and record accurate azimuths, on the horizon, at moonrise and moonset.

It seems likely that the megalithic astronomers knew how to predict eclipses. The purpose of megalithic observations was correction of the lunar count, that is, the numbers of days between possible eclipses. Certainly ability to count was among their skills. Observations of the moon pinpointed the date and position of the moon for the major standstills, and this included observation of the maximum effect of perturbation. It is reasonable to believe these prehistoric skywatchers knew of the perturbation. Observations made 87 days apart would have permitted assessment of its value, at least in terms of its effect on the position of the moon on the horizon. At 14 sites, no perturbation effect appears in the alignment. We reason, therefore, that the mean of observations taken 86 days apart at crest and trough of the perturbation cycle, was used.

People in prehistoric times lived mostly in the open air. Seeing and remembering natural phenomena was part and parcel of their very existence. They were extremely conscious of and responsive to their environment.

Knowledge of the lunar cycle is common to all people. By night, the moon was an important lamp. It is reasonable to imagine that people were able to devise a precise calendar and to predict and pinpoint future aspects of the moon, including eclipses, four thousand years ago. It is also likely that these people would associate the tides with the moon. Safe movement by sea along the shores of continental Europe depended on navigators having an accurate knowledge of tidal ebb and flow, and we know these seaways were used in prehistoric times. Did this association of the moon with the tides prompt further study of the moon's behavior? This is certainly possible, although we do not suggest that the purpose of the lunar observatories and the observations was prediction of tides.

Considerable time and energy were expended on the initial observations. They were what permitted a site to be selected for constitution of the megalithic features. Was the incentive behind this effort scientific curiosity? Were the megalith builders recording their knowledge, in terms of the positions of massive stones, for the sake of their interest in the moon's behavior? Curiosity is, of course, one trait of the human character, and our intellects cannot have altered greatly in the 120 generations that have passed since these prehistoric observatories were erected.

Prehistoric living standards must have been sufficiently high to allow expenditure of the time and resources needed to build the sites and to carry out a long term program of observation. It has been suggested that there were "astronomer-priests" with enough influence to have hundreds of big stones set into their proper positions.

A calendar that reflects the progression of the seasons is a necessity for any organized society--nomadic, savage, or agrarian--and we have evidence to support use of a solar calendar with eight and probably sixteen, "months" per year. Prehistoric astronomer-priests could have pinpointed the times of maximum perturbation at major and minor standstills, and if they did they almost certainly would have noticed the occurrence of eclipses seven days before or after these dates.

No decipherable records remain. The only evidence is the actual positions of stones, markers, or other backsights. Perhaps the cup-and-ring marks, often found on standing stones or on flat rock surfaces, allude to some of these astronomical concerns in a way we have yet to translate.

Our prehistoric ancestors certainly were our equals in intellectual ability. In each generation, there would have been individuals with particular talent. Their not inventing the use of paper and pen does not mean they did not think, speculate, and expound.

How much did the megalithic astronomers really know about the earth, moon, and sun? We firmly believe they knew the moon went around the earth. From high ground on the island of Unst, the most northerly of the Shetlands, the moon at major standstill follows a path that is almost completely above the horizon. The moon is then nearly circumpolar. Some navigators would have sailed farther north than Unst and would have reported that sometimes the moon *never* sets. In the same way, knowledge of the "Land of the Midnight Sun" came south to the peoples of northwest Europe, and we believe megalithic people were capable of reasoning that the sun went around the earth and that the earth was spherical. They could have deduced this last point in two ways: first by noting the earth's shadow on the moon at a lunar eclipse and second by observing the curvature of the sea's surface, a phenomenon obvious to a thoughtful person observing the horizon during a voyage or while climbing a hillside and seeing more distant features come into view.

Only half of all lunar eclipses are visible from any one position on earth, but even so, about 22 eclipses take place every 8.6 years. Did prehistoric record systems include solar eclipses with lunar? We cannot be certain. Fewer solar eclipses will be seen from any given spot, but there is no reason to believe that the *possibility* of a solar eclipse could not have been anticipated. Solar eclipses, like those of the moon, all occur about seven days before or after the perturbation maximum or minimum changing points. It may also have been evident, when the curved limb of the moon was seen against the sun in eclipse, that the agent of obscuration was the moon itself.

Conclusions

What gave the megalith builders their prodigious drive to erect so many dolmens, cairns, mounds, rings, alignments, and other impressive monuments? Over 1500 years from 3000 B.C. to 1500 B.C., they left evidence of their interest in the sun and the moon. We see this in the accurate positions and orientations of many of their monuments, particularly those put up towards the end of the period of megalith building.

With 42 lunar alignments from megalithic sites we have shown that the backsights were products of more than one lunar standstill. It seems likely that the work leading to the erection of a backsight was spread over a period of say, a century and a half. Perhaps this explains why many of the backsights are large stones, erected to last for generations?

The mean obliquity of the ecliptic derived from our 42 sites appears to be about 23°53.1',and this would suggest a date of use and construction near 1590±100 B.C. Our three best solstitial sites provide an obliquity-derived date of 1750±100 B.C.

It now appears certain that by the beginning of the second millennium B.C. megalithic skywatchers observed not only the moon's more obvious behavior but the effects of the 9 arcminute perturbation its orbit experiences as well. Continuous observation would have been undertaken, and eclipses could have been forecast with some success. We are convinced that apart from any practical use the observations may have had, the megalithic astronomers were abstractly interested in the relative movements of the sun, moon, and earth.

We place strong emphasis on the small residuals we encounter when we compare our measured alignments with the nearest expected declinations. Other investigators are often satisfied when such comparisons are within one degree in declination, but we find our residuals so small that their root mean square does not approach 2 arcminutes. The largest residual is about 3 arcminutes, the root mean square about 1.4 arcminutes. Attempts to analyze these results statistically indicate an extremely low probability that the 42 precise lunar alignments are the products of chance. Our interpretation of the intention of their users has followed accordingly.

Reply to Critics

The proliferation of articles, papers, books, and monographs steadily appearing on archaeoastronomy and connecting our name with it is so great that we can not read and assimilate all of it and at the same time continue to do more surveying and analyzing.

Many criticisms of our work are built up on evidence collected by us during the early years of our field surveying. At that time we were feeling our way, and were we now to write a book like Megalithic Sites in Britain, we would, of course, remove from the relevant lists some of the sites thought initially to be astronomically significant. We stand by our latest solar and lunar work seriously attacked by Ellegård in a recent issue of Current Anthropology.

Reply to Ellegård ("Stone Age Science in Britain?" by Alvar Ellegård. Current Anthroplogy, vol. 22, no. 2, April, 1981, pp. 99-125.):

A great deal of what Ellegård writes is directed to his final conclusion that there was no science in bronze age Britain (page 111). He arrives at this decision by simply ignoring everything of ours which appears to disagree with his conclusion. Ellegård attacks the work we have done on the perturbation of the orbit of the moon. He makes much of its very small value. Of course, it is small, and that is why we went back, time after time, to many of the sites to make sure that our measurements were correct. Over the years we have many times plotted histograms, of which those in Megalithic Remains in Britain and Brittany, Fig. 10.9, may be taken as typical. Can anyone look at one of these, understand it, and then say that this is merely the result of chance? L.V. Morrison of The Royal Observatory Greenwich wrote to us and pointed out how much more satisfactory it would be if we divided the histogram into four parts. This is done in Fig. 2 of a paper "A New Study of All Lunar Lines", in Archaeoastronomy (Supplement to J.H.A., no. 2.) In the same journal, Morrison gives a paper on "The Analysis of Megalithic Lunar Lines in Scotland." Anyone really interested in the subject should study this in detail. A.S. Thom has again presented this histogram (1981) with the introduction of "The Lunar Band" (Fig. 1.7).

In a conference held in Oxford in September, 1981, we gave a paper which contained a totally different method of "proof" that the perturbation was observed. We hope Ellegård will read this. At the same conference we hope to give a paper on the 16-month calendar with up-to-date tabulation of the evidence. Ellegård states he does not believe in the calendar. Why?

A great deal of Ellegård's opposition comes from the fact that he tries to take the weather into consideration. This is where we differ. We do not believe that we know what the weather was like in Britain at the beginning of the second millennium. John G. Evans in The Environment of Early Man in the British Isles (Paul Elek, London) discusses climate (page 144). He points out that the bronze age climate was drier than today. Atmospheric conditions could have been very much clearer. We have pointed out how missed observations could be replaced on the ground.

We have, several times, attempted to attach a probability level to our hypothesis. Some of these were, frankly, approximations. But they cannot be far wrong, and every one of them indicated a high degree of probability that we were discovering genuine lunar observatories. We do not mind anyone coming along and saying, "Oh, but--" and then giving a good reason why these probabilities are wrong, but we do not think they should be ignored.

We have gone back to Brogar several times and measured the important foresights again and again. We have been over the ground in detail, and naturally we know much more about the site than can be obtained on a casual visit. For instance did Owen Gingerich, who also criticized our interpretation of Brogar in a commentary appended to the Ellegård paper, really see the ridge on the ground, near the Comet Stone, which we give in Megalithic Remains in Britain and Brittany Fig. 10.7? This shows unequivocally, the notch on Mid Hill, and the mounds along the top of the main ridge at Brogar are also aligned on the notch. There is no argument against this because the notch on Mid Hill is the only notch on the horizon in the neighborhood of the lunar band (see JHA, vi 1975, pp. 100-114, and A.S. Thom, 1981). Did Gingerich, and for that matter Ellegård, consider a point we had made already--namely, that from the center of the main ring the Mid Hill and the cliffs at Hellia were in the correct position for around 3000 B.C.? By the beginning of the second millennium B.C., however, the drop in the obliquity of the ecliptic put these foresights in the wrong position, and, because the foresights could not be moved, the observers proceeded to make the mounds and other features to the south of the ring. This idea suits our measurements perfectly. If one day Gingerich goes back to make his own survey at Brogar, he will see how well mounds A and B provide a suitable stance for a warner whose duty it was to tell the observers that the moon was about to appear, and he will see how Salt Knowe provided a stance for a warner watching for moonrise on Kame of Corrigal. Most of the important lines at Brogar are not obvious on the ground, but they become obvious when the whole site is plotted carefully. The megalithic builders did not leave clues and flags.

C. Ruggles (see below) asks how is it that Thom appears to obtain overwhelming evidence in support of direct observation and recording of the 9 arcminute wobble? The answer is very clear. It is because the alignments are real. The lunar band is a band in the sky which covers all possible positions of the upper or lower limb of the moon at the standstills. We of course are not interested in any foresights which fall outside the lunar band but only with those inside. Were there nothing in our hypothesis, the points on our histogram would be uniformly smeared over the lunar band. Why do we find them piling into little clumps? See any of our histograms, but especially see the one plotted by the method which was suggested to us by L.V. Morrison.

In addition to this there is, of course, a number of foresights which are indicated at the backsight. Perhaps the best is the one to Mid Hill which we have already dealt with in our reply above. One might also mention Dunskeig where the two stones point unequivocally to the foresight.

On page 106 Ellegård briefly discusses the importance to the megalith builders of being able to extrapolate, but we have shown how they did this. The sectors necessary are still on the ground in several places; the extrapolation method which used these has been entirely neglected by Ellegård, who provides no alternative explanation whatever for these sectors.

Ellegård seems not to know of the carved stone balls, of which a large number have been found in Scotland. On these are representations of the five Platonic solids. There are only five of these solids possible and the stones carry all five. (Ref. Time Stands Still by Keith Critchlow, Chap. 7, "Platonic Spheres--a Millennium Before Plato," Gordon Fraser, London, 1979 ISBN 0 86092 0402). It is at present impossible to date these scientific objects. We suggest that they were carefully fashioned by the same people who laid out the stone rings, ellipses, egg shapes, other geometric figures and the cup-and-ring diagrams with such carefully scribed spirals and whorls.

We cannot understand Gingerich's references to the lunar semi-diameter. The values we have used recently are 15.4 arcminutes at the equinoxes and 15.6 arcminutes at the solstices. These we got from L.V. Morrison. Previously we used 15 arcminutes or 15.5 arcminutes. Does Gingerich want us to use two decimals of a minute? Gingerich, like all our other critics, does not pretend to give an alternative explanation for lines of stones, for outliers, for slabs, or for the rows of mounds at Brogar, whereas we have explained most of these with a consistent theory.

Very many years ago we suggested to an eminent astronomer that we should use the word *lunstice*. He immediately turned it down. Our alternative suggestion of "standstill" seems to have caught on, and we can see no point in adopting Ellegård's suggested *lunistice*.

To sum it up, it seems to us that Ellegård began his inquiry not as a neutral and unbiased investigator, but as a complete skeptic. In our opinion, his paper is very misleading, for a casual reading suggests that it is the product of a neutral investigation. In fact, this is not so.

Reply to Ruggles:
("A Critical Examination of the Megalithic Lunar Observatories" by C. Ruggles, foreward by R.J.C. Atkinson. pp. 153-210 in Astronomy and Society in Britain during the Period 4,000 - 1500 B.C., B.A.R. Series 88, 1981. Edited by C.L.N. Ruggles and A.W.R. Whittle, and "Prehistoric Astronomy: How Far Did It Go?" by C. Ruggles, New Scientist, 18 June 1981, pp 750 - 753.)

We were very glad to see that at last someone who can really use a theodolite was going over the various sites we have measured and interpreted. As we cannot go over them again ourselves, we must accept some of Ruggle's corrections, but we cannot agree with many of his conclusions and much of his harsh critique. He has based his conclusions on an incomplete sample, and , it seems to us, without any sympathy for the subject or, in fact, with the erectors. Let us look at a few lines. Why does he discard the alignment at Kell Burn? This was an alignment of small stones with a rectangle or square nearby. Why is he so critical of the Glen Prosen alignment? It is true that the horizon is now featureless, but any error made in estimating the azimuth of the alignment itself would have very little effect on the declination.

We agree with some of his remarks about Haggstone Moor and especially with Atkinson's remarks about stone *S*, but why does Ruggles discard the profile measured from Long Tom? The obvious foresight is the gap behind Torr More. As far as we know this has never been measured from the Haggstone itself. Why did he not do this?

At Mid Clyth he seems to have missed the fallen stone menhir to the west, and why does he discard our measurements of the hills across the Moray Firth (Megalithic Lunar Observatories, Fig. 9.2)? He also missed the alignment of the row of boulders on top of the hill, which form, in our opinion, a position for a man to stand to give warning that the moon was coming up on the small foresight.

We see no reason why, as Ruggles suggests, the erectors should have duplicated solstitial lines in the vicinity of Ballochroy. The hard work involved in fixing the site had been done once, and to repeat it would not have been necessary.

Why does Ruggles dismiss Brogar the way he does? Is it because the backsights are mounds? We believe that all of the observing positions were moved from the ring over the centuries. Probably the move was not complete, and they had not time to put in the stones, except of course the Comet Stone. We do not see why stones only should have been used as backsights. The mounds make perfectly good indicators. Ruggles says nothing at all about the three warning positions--at cairn A, cairn M, and Salt Knowe. He complains that we looked around Stonehenge for other foresights to suit the various positions. Why should we not? It is easy to look at a map and point out various features on the map and decide how many of these there are. But how many of them are on the horizon of Stonehenge? This is the crucial point. Gibbet Knoll and Figsbury Ring are, I reckon, on the horizon, but apparently Atkinson does not want us to use these two. The hole and mound (Peter's Mound) which are on the horizon from Stonehenge in exactly the correct position for the solstitial rising sun, we are told now are modern. Peter Newham was told by an old man that there had been a stone buried there, but it would not form a suitable foresight at that distance. But another and more important point is that the Army surveyors told us that they had looked over the various plans of huts and other structures which had been erected on this plateau over the years, and the position of Peter's mound has always been kept clear. What was it?

So far Ruggles seems to have surveyed only one sector, that at Loch of Yarrows. He chose, perhaps, the worst to begin with. What about Mid Clyth, Dirlot and Camster? He has not compared our Dirlot survey with that made by Dryden last century, nor is he prepared to make any use of the profile we measured from the cairn at the top of Dirlot rows (Megalithic Lunar Observatories, Fig. 9.5)

He dismisses many of our proposed sites because there is no indication of the foresight or backsight. Does he expect all the small stones in a field to remain for thousands of years? The farmer probably only left the large one. He dismisses Campbeltown, but we are convinced that the erectors were really lucky to find on the meridian a foresight as accurate as that we show. This is the only case known to us where the moon is recorded crossing the meridian, and, of course, it is at an altitude of about 5°. And so refraction can be considered to be known accurately.

How does Ruggles explain the fact that on his survey of Quinish the stone is not oriented along his proposed line but more-or-less along our proposed line? He says nothing about the fact that the alignment of large menhirs at Escart Farm seems to serve two purposes: (1) the general line to Sheirdrim Hill where the moon sets at major standstill; (2) the two stones at each end of the alignment, which indicate the solstice sunset. And so it goes on. His criticisms are nearly always destructive. Nowhere has he made an attempt in the field to support our ideas. To the casual reader he gives the impression that we cannot survey and that all our astronomical work is completely worthless. Ruggles asks what would we have found from the Brogar histogram (Megalithic Remains in Britain and Brittany) had we considered declinations some 10 arcminutes different, which would require about 1,300 years' change in date. We considered seriously, many years ago, the possibility of moving the humps along to the next position on the histogram, but because this involved a change of over 1,000 years, we dropped the idea completely. It made the date much too early for the type of careful observations we believe the megalithic astronomers were making.

Reply to J. Patrick("A Reassessment of the Lunar Observatory Hypothesis for the Kilmartin Stones" by J. Patrick.) Archaeoastronomy, Supplement to JHA, no. 1, pp. S78-S85.

We firmly believe that Temple Wood is a lunar observatory, and we look upon Patrick's survey as an excellent check of our survey work. The foresight is at a distance of 2 km, and the azimuth, depends crucially upon where the surveyor's theodolite is placed. Vegetation 2 meters deep, and/or trees, would, of course, blur or obscure any horizon notch. The observers, however, neither would have had difficulty in keeping the notch clear, nor would they themselves have experienced any difficulty in knowing which notch they were using. After all, they were living there. They watched the skies continuously and could see when the moon was approaching its greatest azimuth. Any grass growing on the notch would not, in our opinion, remove the notch's effectiveness.

Also, Patrick asks, why megalithic astronomers would build a similar observatory at Barbreck, only 5 miles away from Temple Wood? Two observing teams would have been necessary. Two stones near the house stand at a small angle to each other, and it is possible that they are the remains of a large circle, now built over by the house and barns. There are also several stones to the north of Barbreck site which might be the remains of an observatory. Patrick does not mention them in his analysis.

Appendix Table 1.

42 Lunar Lines

Site	δ_o	Nominal δ	Ω for $c_s = 0$	ε 23°	β	R $= \lvert\beta - \Omega\rvert$	Ω for $c_s = 0.6'$ or $c_s = 0.3'$	R $= \lvert\beta - \Omega\rvert$	Ref.
Brogar JK	−18°44.1'	$-(\varepsilon-i)$	−0.7'	52.1'	0.3'	1.0'	−1.0'	1.3'	3
Brogar M	−18°59.2'	$-(\varepsilon-i+s)$	−16.2	51.7	−14.8	1.4	−16.5	1.7	3
Brogar Comet	−18°50.3'	$-(\varepsilon-i-\Delta+s)$	−6.8	52.2	−5.9	.9	−7.4	1.5	3
Brogar MLJ	−18°20.1'	$-(\varepsilon-i-\Delta-s)$	24.0	52.8	24.3	.3	23.4	.9	3
Brogar Kame l	29°24.3'	$+(\varepsilon+i+\Delta+s)$	24.0	51.6	22.5	1.5	23.4	.9	3
Brogar Kame u	29°24.8'	$+(\varepsilon+i+\Delta+s)$	24.0	52.1	23.0	1.0	23.4	.4	3
Brogar L	−18°50.9'	$-(\varepsilon-i-\Delta+s)$	−6.8	52.8	−6.5	.3	−7.4	.9	3
Brogar Ravie Hill, Comet	28°48.2'	$+(\varepsilon+i-s)$	−16.2	55.7	−13.6	2.6	−16.5	2.9	3
Brogar Ravie Hill, H	28°52.3'	$+(\varepsilon+i+\Delta-s)$	−6.8	50.4	−9.5	2.7	−7.4	2.1	3
Stenness	19°08.0'	$+(\varepsilon-i+\Delta+s)$	25.6	51.1	23.6	2.0	26.2	2.6	3
Callanish I	−29°12.8'	$-(\varepsilon+i+\Delta)$	−8.6	55.5	−11.0	2.4	−8.0	3.0	2 H1/1
Callanish V	−28°35.4'	$-(\varepsilon+i-\Delta-s)$	25.6	52.3	26.4	.8	26.2	.2	2 H1/5
Mid Clyth A_1	−29°09.2'	$-(\varepsilon+i+\Delta)$	−8.6	51.9	−7.4	1.2	−8.0	.6	2 N1/1
Mid Clyth A_2	−29°27.1'	$-(\varepsilon+i+\Delta+s)$	−24.0	54.4	−25.3	1.3	−23.4	1.9	2 N1/1
Wormadale Hill	−29°09.5'	$-(\varepsilon+i-\Delta+s)$	−8.6	52.2	−7.7	.9	−8.0	.3	14
Ballinaby	29°16.8'	$+(\varepsilon+i+s)$	14.8	53.3	15.0	.2	14.5	.5	15
Ballinaby	29°25.6'	$+(\varepsilon+i+\Delta+s)$	24.0	52.9	23.8	.2	23.4	.4	3

Appendix Table 1, continued

Site	δ_o	Nominal δ	Q for $c_s = 0$	ε 23°	β	R $= \lvert\beta - Q\rvert$	Q for $c_s = 0.6'$ or $c_s = 0.3'$	R $= \lvert\beta - Q\rvert$	Ref.
Kintraw	-18°20.9'	$-(\varepsilon-i-\Delta-s)$	24.0'	53.6'	23.5'	.5'	23.4'	.1'	2 A2/5
Temple Wood S_1A_1	28°55.4'	$+(\varepsilon+i+\Delta-s)$	-6.8	53.5	-6.4	.4	-7.4	1.0	2 A2/8
Temple Wood $S_{54}A_1$	29°18.1'	$+(\varepsilon+i+s)$	14.8	54.6	16.3	1.5	14.5	1.8	2 A2/8
Temple Wood Q A_1	29°01.5'	$+(\varepsilon+i)$	-0.7	53.5	-0.3	.4	-1.0	.7	2 A2/8
Temple Wood A_2	-28°48.3'	$-(\varepsilon+i-s)$	16.2	55.8	13.5	2.7	16.5	3.0	2 A2/8
Ballymeanach A_1	29°12.8'	$+(\varepsilon+i+\Delta)$	8.6	55.5	11.0	2.4	8.0	3.0	2 A2/12
Ballymeanach A_2	29°27.7'	$+(\varepsilon+i+\Delta+s)$	24.0	55.0	25.9	1.9	23.4	2.5	2 A2/12
Dunadd	29°16.4'	$+(\varepsilon+i+s)$	14.8	52.9	14.6	.2	14.5	.1	2 A2/13
High Park A_1	-29°27.7'	$-(\varepsilon+i+\Delta+s)$	-24.0	55.0	-25.9	1.9	-23.4	2.5	2 A4/2
High Park A_2	-28°46.3'	$-(\varepsilon+i-s)$	16.2	53.8	15.5	.7	16.5	1.0	2 A4/2
Skipness	-29°24.4'	$-(\varepsilon+i+\Delta+s)$	-24.0	51.7	-22.6	1.4	-23.4	.8	16
Beacharr	29°11.1'	$+(\varepsilon+i-\Delta+s)$	8.6	53.8	9.3	.7	8.0	1.3	2 A4/5
Crois Mhic-Aoida	-28°37.2'	$-(\varepsilon+i-\Delta-s)$	25.6	54.1	24.6	1.0	26.2	1.6	2 A4/9
Campbelltown	-28°44.8'	$-(\varepsilon+i-s)$	16.2	52.3	17.0	.8	16.5	.5	2 A4/14
Campbelltown	-28°54.9'	$-(\varepsilon+i+\Delta-s)$	6.8	53.0	6.8	0.0	7.4	.6	2 A4/14
Knockstaple	28°52.5'	$+(\varepsilon+i+\Delta-s)$	-6.8	50.6	-9.3	2.5	-7.4	1.9	2 A4/19
Gigha	29°25.7'	$+(\varepsilon+i+\Delta+s)$	24.0	53.0	23.9	.1	23.4	.5	2 A4/17
Dunskeig	-19°00.0'	$-(\varepsilon-i+s)$	-16.2	52.5	-15.6	.6	-16.5	.9	6
Knockrome	-29°27.0'	$-(\varepsilon+i+\Delta+s)$	-24.0	54.3	-25.2	1.2	-23.4	1.8	2 A6/4
Stillaig	28°53.4'	$+(\varepsilon+i+\Delta-s)$	-6.8	51.5	-8.4	1.6	-7.4	1.0	2 A10/5
Glen Prosen	-28°55.1'	$-(\varepsilon+i+\Delta-s)$	6.8	53.2	6.7	.1	7.4	.7	2 P3/1

Appendix Table 1, continued

Site	δ_o	Nominal δ	Q for $c_s = 0$	ε 23°	β	R $=\lvert\beta - Q\rvert$	Q for $c_s = 0.6'$ or $c_s = 0.3'$	R $=\lvert\beta - Q\rvert$	Ref.
Lundin Links	-18°59.7'	$-(\varepsilon-i+s)$	-16.2'	52.2'	-15.3'	.9'	-16.5'	1.2'	2 P4/1
Fowlis Wester	29°17.0'	$+(\varepsilon+i+s)$	14.8	53.5	15.2	.4	14.5	.7	2 P1/10
Parc y Meirw	18°19.6'	$+(\varepsilon-i-\Delta-s)$	-24.0	52.3	-24.8	.8	-23.4	1.4	2 W9/7
Blakely Moss	28°45.8'	$+(\varepsilon+i-s)$	-16.2	53.3	-16.0	.2	-16.5	.5	2 L1/16

NOTE: The values of ε and β are deduced from δ_o, the observed declination for the mean of day and dark.

$\beta = \delta_o - (\varepsilon \pm i)$

$Q = \delta_e - (\varepsilon \pm i)$ = expected value of $\beta = \pm(\Delta \pm s)$

where δ_e = nominal or expected declination

R = residual = $\beta - Q$,
irrespective of sign.

r = root mean square of R

For example, ε for the major standstill and for positive declination is given by $\varepsilon = \delta_o - i - Q$.

The mean i = 5° 8.7'

From above 42 values of β, r equals 1.34' for assumption 1 and 1.53' for assumption 2. In assumption 1, the observers determined their position over a span of about 179 years. For this, c_s would have been 0. (See first seven columns.)

In assumption 2, the observers had determined the position of the stone from one standstill. For this, c_s may be 0.3' or 0.6' (columns 8 and 9).

Appendix Table 2.

Evolution of the Analysis of Lunar Lines.

By Year	New Lines	Total Lines to Date	Unacceptable Lines	Total of Lines Discarded to Date	Lines Used in Ref. 1, Ref. 2, or Here	Notes
1967	37	37	—	—	37	Ref. 1 Fig. 10.1
1971	43	80	28	28	52	Ref. 2 Tables 7.1, 7.2
1980	15	95	25	53	42	Here

Appendix Table 3.

List of 25 lunar lines discussed in Ref. 2 with reasons why each line is now unacceptable.

Site	Name	Reason
H3/11	Leacach and Tighe Chloiche A_2	Map work only
H1/1	Callanish I	Map work only; Unsure of hill behind the hollow
M1/3	Quinish A	Survey too hurried; no sun.
H1/5	Callanish V; to N.W.	Position of backsight unknown
H1/5	Callanish V A_2 Fig. 6.14	Only foresight A, admissable because stones indicate Mor Monach
H1/5	Callanish V A_4 Fig. 6.14	" " " "
		" " " "
		" " " "
		" " " "
N1/17	Dirlot A_1 Fig. 9.5	Unsure of backsight
N1/17	Dirlot A_2	" " " "
H3/11	Leacach an Tighe Cloiche A_1	See above
A4/1	Escart A; Fig. 6.1	Survey uncertain; trees
A2/5	Kintraw A_2. Fig. 4.4	Ambiguous notch
G3/2	Haggstone Moor. Fig. 6.17	Uncertainty of backsight
A4/9	Crois Mhic-Aoidha; Fig. 5.6	Line uncertain
G3/2	Haggstone Moor. Fig. 6.17. A_1	See above
A4/17	Gigha Fig. 6.5; A_1	Foresight uncertain

Appendix Table 3, continued

Site	Name	Reason
P4/1	Lundin Links; page 56, Ref. 2	Map work only for Cormie Hill
A2/13	Dunadd B. Fig. 6.8	No indication
A6/1	Camus an Stacca Fig. 6.9	Too many notches; uncertain
A6/1	Camus an Stacca Fig. 6.9	Too many notches; uncertain
M1/3	Quinish	See above
G3/2	Haggstone Moor Fig. 6.17	See above
G3/2	Haggstone Moor Fig. 6.17	See above
P1/10	Fowlis Wester B Fig. 5.4	No foresight on near horizon.
A2/8	Temple Wood Notch B	Azimuth uncertain surveying accur- acy marred by two planations
G9/131	Kell Burn p. 70 Ref. 2	Short line; no foresight

References

MacKie, E.W.
1977 Science and Society in Prehistoric Britain. London: Elek Books Limited.

MacKie, E.W. and J.S. Bibby
1974 "Archaeological Tests on Supposed Prehistoric Astronomical Sites in Scotland," Philosophical Transactions of the Royal Society, vol. 276, no. 1257 ("The Place of Astronomy in the Ancient World," ed. F.R. Hodson), pp. 169-194.

Morris, R.W.B.
1977 Prehistoric Rock Art of Argyll. Poole, England: Dolphin Press.

Morris, R.W.B.
1979 Prehistoric Rock Art of Galloway and the Isle of Man: Blandford.

Renfrew, C.
1979 Investigations in Orkney. London: Society of Antiquaries (Distributed by Thames and Hudson Ltd.)

Ritchie, J.N.G.
1975-76 "The Stones of Stenness, Orkney," Proceedings of the Society of Antiquaries of Scotland, vol. 107, p. 50.

Thom, A.
1967 Megalithic Sites in Britain. Oxford: Oxford University Press.

Thom, A.
1971 Megalithic Lunar Observatories. Oxford: Oxford University Press.

Thom, A.
1974 "A Megalithic Lunar Observatory in Argyll," Journal for the History of Astronomy, vol. 5, pp. 50-51.

Thom, A. and R.L. Merritt
1978 "Some Megalithic Sites in Shetland," Journal for the History of Astronomy, vol. 9, pp. 54-60.

Thom, A. and A.S. Thom
1971 "The Astronomical Significance of the Large Carnal Menhirs," Journal for the History of Astronomy, vol II, pp. 147-160.

Thom, A. and A.S. Thom
1978 Megalithic Remains in Britain and Brittany. Oxford: Oxford University Press.

Thom, A. and A.S. Thom
1978 "A Reconsideration of the Lunar Sites in Britain," Journal for the History of Astronomy, vol. 9, pp. 170-179.

Thom, A. and A.S. Thom
1979 "Another Lunar Site in Kintyre," Archaeo-astronomy (supplement to Journal for the History of Astronomy), No. 1, pp. S96-S97.

Thom, A. and A.S. Thom
1979 "The Standing Stones in Argyllshire," Glasgow Archaeological Journal, vol. 6, pp. 5-10.

Thom, A. and A.S. Thom
1980 "Astronomical Foresights used by Megalithic Man," Archaeoastronomy (supplement to Journal for the History of Astronomy, vol. 11), No. 2, pp. S90-S94.

Thom, A. and A.S. Thom
1980 "A New Study of All Megalithic Lunar Lines," Archaeoastronomy (supplement to Journal for the History of Astronomy), No. 2, pp. S78-S89.

Wood, J.E.
1978 Sun, Moon and Standing Stones. Oxford: Oxford University Press.

Ronald Hicks

4. Stones and Henges: Megalithic Astronomy Reviewed

It was first suggested over two hundred years ago that Stonehenge served an astronomical--or at least calendrical--purpose. Sir Norman Lockyer (1909) pursued this theme in more detail at the beginning of this century, and in the following years Admiral Boyle Sommerville (1912, 1923) extended astronomical studies to Callanish in the Hebrides and other stone circles. But it was not until the mid-1960 s, as an outcome of the work of C.A. Newham (1964, 1972) and Gerald Hawkins (1965) at Stonehenge and of Alexander Thom (1967, 1971) at a large number of sites in Scotland and Wales, that megalithic astronomy could really be said to have come of age. It was these studies that caught the public imagination and forced archaeologists and others interested in megalithic sites to make a serious attempt to come to grips with the problem of the purpose and meaning of these monuments and the question of what they can tell us about the knowledge and sophistication of their builders. The resulting work has included not only studies of the possible astronomical uses of the monuments but also of the system of measurements used in their planning, the geometrical knowledge evident in their layouts, iconography as reflected in accompanying petroglyphs, and the type(s) of social organization necessary to account for the sometimes prodigious labor that went into their construction. It is timely, therefore, to review the accomplishments to date, the controversies that have arisen, and the problems that remain.

The studies carried out by Alexander Thom, with the assistance of his family and other colla-

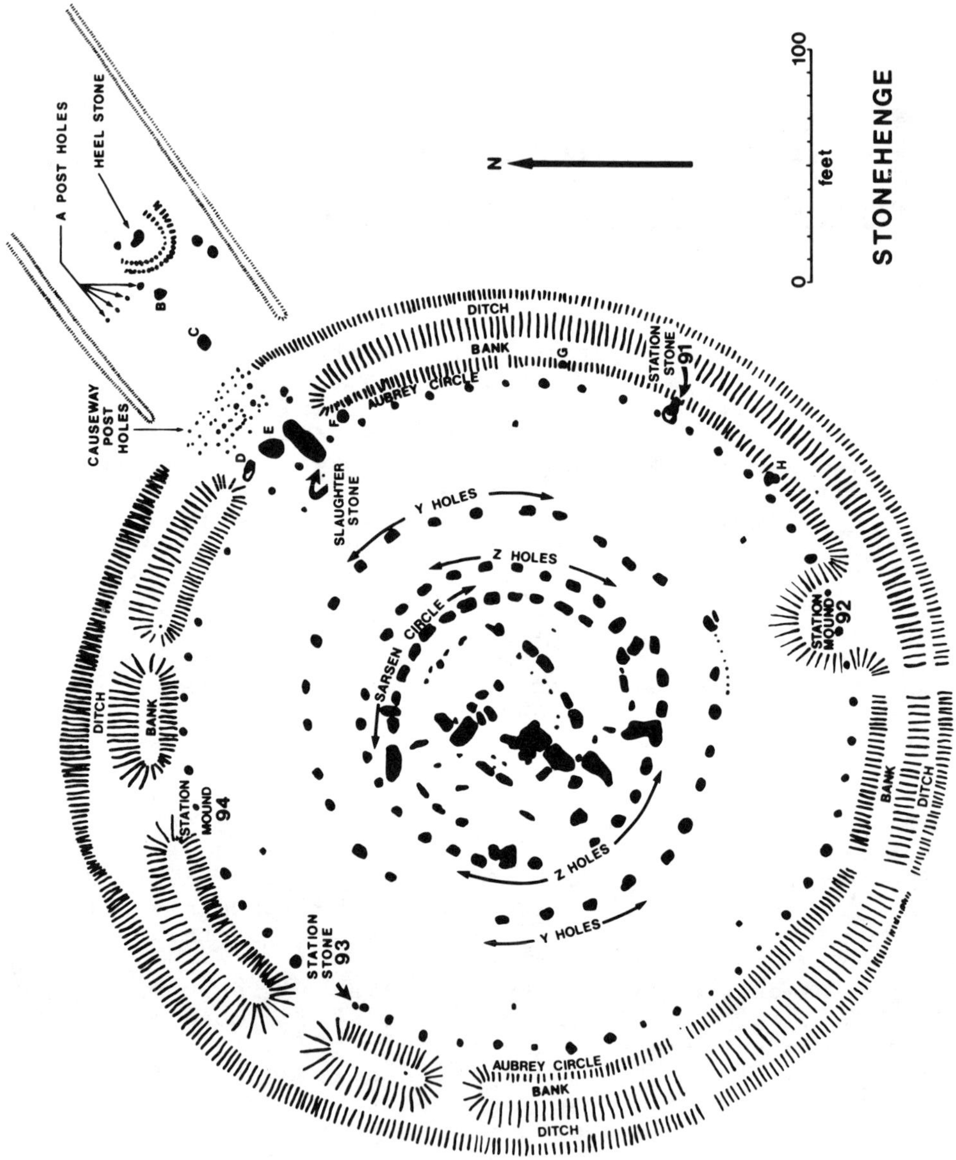
A POST HOLES
HEEL STONE
B
C
CAUSEWAY POST HOLES
D
E
F
G
H
SLAUGHTER STONE
DITCH
BANK
AUBREY CIRCLE
STATION STONE 91
STATION MOUND 92
STATION STONE 93
STATION MOUND 94
Y HOLES
Z HOLES
SARSEN CIRCLE
Z HOLES
Y HOLES
AUBREY CIRCLE
BANK
DITCH
N
0
feet
100
STONEHENGE

With an axis of symmetry oriented northeast-southwest, Stonehenge has been associated with the summer solstice sun. Other alignments between various features noted here have been linked with other significant solar events and with the inner and outer limits of the moon's monthly extremes during the 18.6 year "standstill" cycle. New discoveries do not necessarily clarify the meaning of these alignments. A prehistoric pit that once contained a large stone was discovered in 1979 near the Heel Stone, popularly regarded as a sunrise marker. The relationship between these features remains ambiguous. (Griffith Observatory) borators, have clearly dominated the astronomical, metrological, and geometrical interpretations of the megalithic monuments of the British Isles and neighboring Brittany. The core of Thom's work has been collected and restated in three books--Megalithic Sites in Britain (1967), Megalithic Lunar Observatories (1971), and Megalithic Remains in Britain and Brittany (Thom & Thom, 1978). The first of these summarizes Thom's methods and basic hypotheses, while the other two elaborate various aspects of his work and consider particular important sites in greater detail. Because this body of work is important, unavoidably it will be central to much of what I have to say. I do not wish to become overly involved in the analysis of specific sites, however, and so for the most part I shall discuss, in general, the approaches, the scope of the work carried out, and the conclusions. Individual sites will be mentioned only where it seems necessary to clarify points or document the disagreements between various researchers.

Astronomical and Calendrical Interpretations

The idea that the stone circles of the British Isles might be intended, at least partly, for astronomical observations first arose when William Stukeley in 1740, followed by Dr. John Smith in 1771, noted that the summer solstice sunrise can be observed by standing at the center of Stonehenge and looking down the avenue leading from the monument's entrance. In 1846 the Rev. Edward Duke recognized that the so-called Station Stones rectangle--a feature belonging to late in the first phase of construction and now marked by two stones and two low mounds where two other stones once stood, numbered clockwise as 91 through 94--also were astronomically significant. The view to stone 91 from mound 92 coincides with summer solstice sunrise. Looking across stone 93 from mound 94, one can observe winter solstice sunset (Newall 1959:20-21). Equinox sunrise is indicated by looking from mound 94 to the site of a now missing stone near the entrance to the avenue, a fact first noted only in 1963 by C.A. Newham. Further, Newham also observed that the mound 94 to stone 91 alignment marks the most southerly moonrise. Similarly, the most northerly moonset is given by the mound 92 to stone 93 alignment. About the same time Newham was carrying out his work, Gerald S. Hawkins was en-

From the center of Stonehenge, the Heel Stone is framed by arches of the Sarsen Circle. The arches are part of the third phase of Stonehenge, and the Heel Stone belongs to the first. The Avenue, which continues northeast from the vicinity of the Heel Stone was constructed in Stonehenge's second stage. The monument's history is, therefore, complex. (E.C. Krupp)

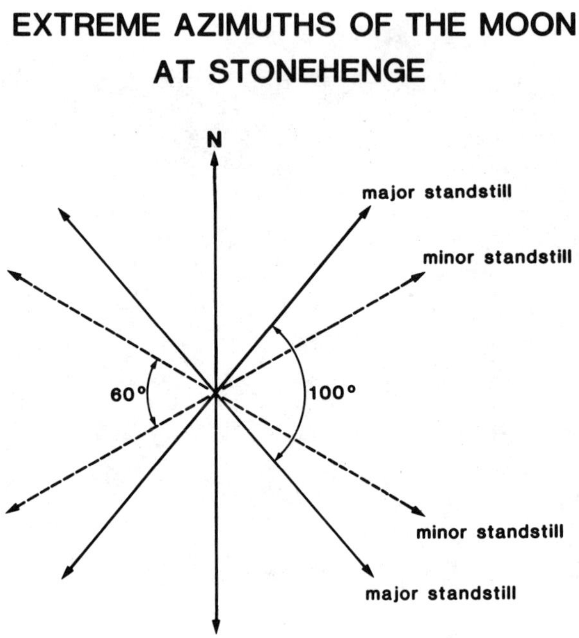

At the latitude of Stonehenge (about 51° north), the change in the month's moonrise and moonset limits is dramatic over the full 18.6-year standstill cycle. At major standstill the northernmost moon rises about 20 degrees farther north than the northernmost moon at the minor standstill, 9.3 years later. This excursion of the moon is very noticeable. (Griffith Observatory)

gaged in an even more ambitious task, calculating the declinations for some 120 possible alignments at Stonehenge, taking into account the center, the trilithon archways, stones, stoneholes, other holes and mounds. The initial effect of Hawkins's work was to confirm that the solar and lunar extremes were reflected repeatedly in the Stonehenge alignments. He identified (1973:61) ten solstice alignments, a like number for the extreme lunar rising and setting positions, and four for what Thom has called the "minor lunar standstill." This last is also an extreme lunar position, because apparent lunar motion is considerably more complex than that of the sun. The moon's monthly swing shifts from an outer (major) to an inner (minor) extreme declination and back again over a period of 18.61 years.

Thus by the mid-1960s, a number of the alignments essential to an understanding of ancient astronomy in the British Isles had been recognized at Stonehenge, and it was clear that the monument could have been used to establish a rudimentary solar calendar, rather like a giant sundial, marking the seasons instead of the hours. Alignments marking the lunar extremes opened up another possibility as well. Stonehenge could have served as an eclipse predictor. Whether in fact it did, and if so, exactly how, is a point on which controversy has continued. Sir Fred Hoyle (1966, 1977) argued that indeed it was an eclipse predictor, but not in quite the way Hawkins imagined.

Evidence for an intense interest in astronomy at Stonehenge, some four to five millennia ago, looks convincing. How much more convincing it would be if we could find similar evidence in other monuments of the period. Here the work of Alexander Thom is clearly important. Over 900 stone circles are known to have existed in the British Isles (Burl 1976a:8), and by the time Hawkins's work appeared, Thom had spent over 30 years carrying out painstakingly accurate surveys of dozens of these monuments. His analyses produced results very similar to those found at Stonehenge, as well as some additional information.

The Calendar

In 1901 Sir Norman Lockyer had suggested that intermediate solar positions also were incorporated

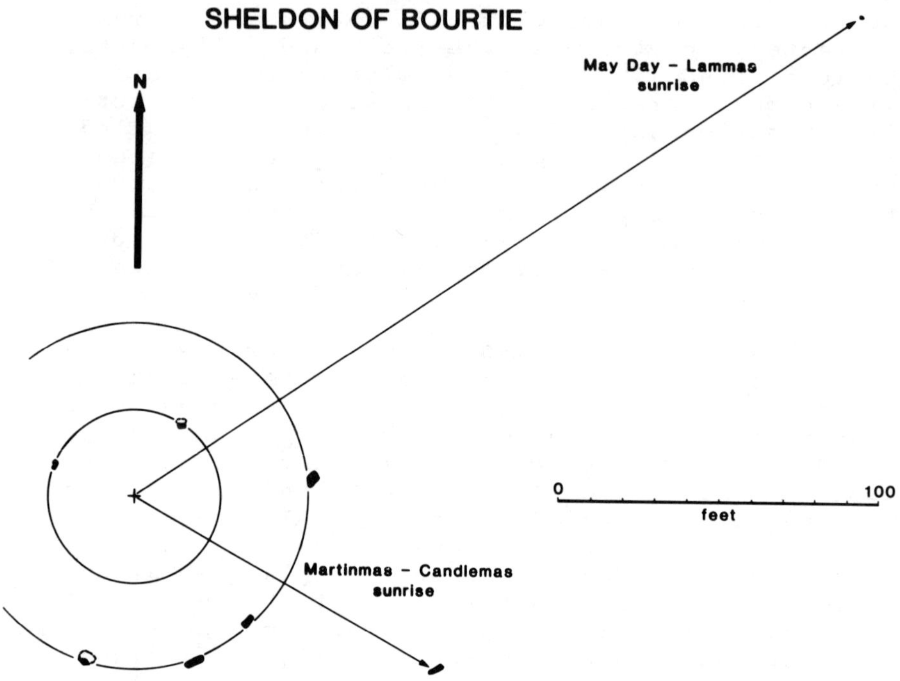

Sunrises and sunsets on key dates besides the solstices and equinoxes may have been incorporated into megalithic monuments. Here at Sheldon of Bourtie, a pair of concentric stone circles in Aberdeenshire, Scotland, two outlying stones combine with the circles' common center to form alignments with the sunrise for traditional holidays--May Day/Lammas and Martinmas/Candlemas--that correspond to old Celtic festivals and the old "quarter days." (after A. Thom, A.S. Thom, and A. Burl, Griffith Observatory)

into Stonehenge to divide the year into eighths. Boyle Somerville found similar alignments in sites he investigated a decade or so later. A single alignment can allow observation of sunrise near the beginnings of May and August and, if used in reverse with the proper horizon, near the beginnings of February and November. These are near the dates of the old quarter days, the first days of the seasons in the pre-Christian tradition of the British Isles. On these days, festivals--Imbolc, Beltaine, Lughnasa, and Samhain--were celebrated. In Christian tradition, they are equivalent to Candlemas, May Day, Lammas, and Martinmas, and so we certainly have evidence of their having been important in early historical times. One of the earliest recorded traditions concerning Stonehenge declares that it was built to commemorate a May Day massacre. Other stone circles also have May Day or Lughnasa traditions.

Thom's surveys confirmed the existence not only of quarterday alignments but also, in his opinion, of another set, marking declinations near $\pm 22^{\circ}$, $+9^{\circ}$, and -8°. These would have allowed division of the year into sixteen "months," each from 22 to 24 days long. I have found a number of orientations near these last declinations in Irish sites, but I don't feel they are quite as unambiguous as those for other dates. A strong argument can be made for the existence of a protohistoric solar-lunar calendar in Britain, a combined calendar which required the periodic addition of an intercalary lunar month to keep the solar and lunar components more-or-less synchronized over the nineteen-year period necessary for solar and lunar dates to correlate exactly. Two millennia separate the megalithic and protohistoric periods, however, and calendars do change. Whether the evidence best supports a 16-month solar calendar or a combined solar/lunar calendar is unclear, and more work needs to be done.

Megalithic Lunar Observatories

Publication of *Megalithic Lunar Observatories* in 1971 indicated Professor Thom's view of megalithic astronomy had evolved. Nearly all the impressive alignments in Britain, he now felt, were lunar. In Thom's view great importance was attached to these sites by their builders. Indeed, he argued they had carried their study of the moon's movements to a precision that would have enabled them to predict which full or new moons would coincide with eclipses. In 1971 he considered two sites most important for an understanding of megalithic lunar observatories in Britain: Temple Wood (otherwise known as the Kilmartin Stones) in Argyllshire and Mid Clyth in Caithness (Thom 1971:45).

Superimposed on the moon's more obvious cycles is a perturbation cycle of ± 9 arcminutes with a period of 173.3 days (one half an eclipse year). This produces small irregularities in the lunar standstill positions, and Thom believed that it would have been possible to observe this perturbation and to carry out the extrapolations needed to locate the actual standstill position, should it fall between two moonrises. This may have been required to set up the observatory. In more recent work, Thom and his son, A.S. Thom, have argued for a similar use of the stone rows of Brittany (Thom & Thom 1971, 1972, 1978; Thom, Thom, & Gorrie 1976) and have called the Brogar complex in Orkney the most complete megalithic lunar observatory surviving in Britain (Thom & Thom 1978).

The work of the Thoms might seem to have answered adequately whether megalithic astronomers were sufficiently knowledgeable to have predicted eclipses. But their studies, like those postulating eclipse prediction at Stonehenge, prompted a variety of criticisms, from D.C. Heggie (1972), J. Patrick & C.J. Butler (1974, Patrick 1976), Evan Hadingham (1981), and others. Gordon Moir (1979) has carefully reviewed all the difficulties involved in making the necessary precise observations. Moir directed his remarks specifically at Stonehenge but noted that many of them would apply to the other sites as well. He has concluded that the major and minor lunar standstill positions could have been identified in a general way, but he doubts the precision necessary for eclipse prediction was attainable.

It is not possible to establish, from the stone alignments alone, that the builders of the monuments knew anything more about the moon's motion than the 18.6-year cycle of oscillation be-

The Ring of Brogar, in Orkney, is, according to the Thoms, one component of a complex prehistoric lunar observatory. Here, looking southeast, the distinctively shaped stone slabs form a circle, which stands upon a circular "platform" of earth, bounded by a ditch, visible in the foreground. (Robin Rector Krupp)

tween the major and minor standstill positions. For more exact observations, a foresight on the distant horizon was required, but it is frequently unclear which of two or more possible foresights was intended. Sometimes it is uncertain exactly where the observer was to stand. There are several other difficulties: 1) The extreme declinations are not likely to occur exactly at moonrise or moonset when the observations are being made; 2) The apparent rising and setting points of the moon are affected by refraction, which is in turn temperature-dependent; 3) Observation of moonrise or moonset is in any case difficult in daytime; 4) It becomes doubly so if the moon is in any phase but full, which it will be more often than not; and 5) There is always the likelihood of clouds or haze to prevent observation at the time of the extreme standstill positions, which occur only at 9.3-year intervals.

While the Thoms have attempted to compensate for these problems, when all these complications are taken into account, I personally feel that the observed sightlines are most likely to represent the best that could be done at the time of any given standstill, with any coincidence with the lines required for precise observation of a perturbation being purely coincidental. Nonetheless the controversy over whether megalithic astronomers could or could not predict eclipses is likely to continue.

Planets and Stars

So far no mention has been made of the place of the planets and stars in megalithic astronomy. For the moment this must remain the case for the planets, for no one has been able to demonstrate any clear connection between the observed alignments and planetary movements, although Venus may explain a cluster of Irish alignments on declination -32.5° (Barber 1973:34), as well as a few sites in Scotland (Burl 1976:177). It seems highly unlikely that the megalith builders would have ignored the planets completely. This, then, is yet another area in which further research may reward us.

The stars have fared somewhat better, but the

gradual shift in their positions leads to difficulties. Most of the megalithic monuments have not been accurately dated, and star alignments are date dependent. Other problems are discussed by Barber (1973:34-35) and Burl (1976a:51-54). In Megalithic Sites in Britain, Thom assumed a range of dates from 2100 to 1500 B.C. During this six-century span sixteen first magnitude stars would have been visible in Britain (only Sirius at the level of the horizon). Thom concluded that only Capella was widely observed. Altair and Pollux may have been used also, and probably all served as timekeepers, at various sites. The declination of Regulus agreed with the declination of the summer solstice sun, and Aldebaran's declination coincided with the equinoctial sun, so it cannot be determined whether Regulus and Aldebaran were used as well. It is curious that Sirius, the brightest of the stars, was not favored with any known alignments. If an earlier date is assumed for the sites, as seems to be required by recalibration of the radiocarbon dates (Thom & Thom 1978:5), then the case for Capella is weakened and only Deneb stands out. Certainly the stars, like the planets, need more study. The Pleiades and other groups are well worth examining more closely, as well.

Passage Graves, Henges, and Cursus Monuments

To this point we have considered results derived largely from the study of stone circles and alignments, but other megalithic monuments--and some related monuments not involving the use of big stones--have also contributed to our knowledge of ancient astronomy.

Passage graves are monumental artificial mounds composed of earth or small stones within which a passageway of megaliths leads to a stone chamber. Two of the most impressive passage graves still surviving in the British Isles--Newgrange in Ireland and Maes Howe in Orkney--are aligned with the midwinter solstice sun. Its rays shine down the passages and illuminate the inner chambers. At Newgrange where the passage is low, narrow, and 70 feet long, the effect is dramatic and spectacular. An astronomical explanation does not, however, work for the smaller passage graves, for example, satellite monuments surrounding another member of

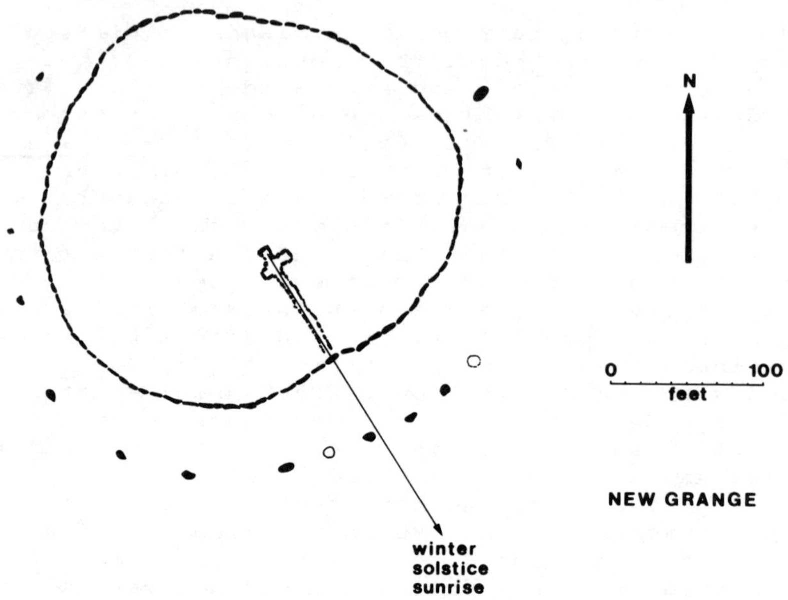

Newgrange has an axis of symmetry that is oriented toward the southeast and coincides reasonably well with the direction of winter solstice sunrise. (Griffith Observatory)

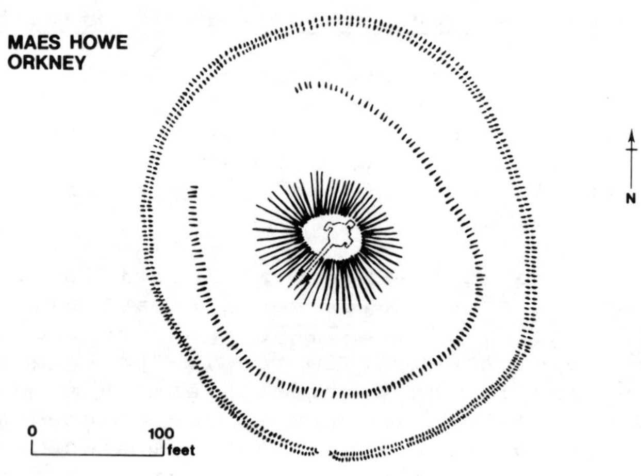

Maes Howe, an elegantly constructed neolithic passage grave in Orkney, not far from the Ring of Brogar, has its corridor and entrance oriented toward the southwest and winter solstice sunset. (Griffith Observatory)

the Newgrange group, Knowth. These are oriented toward the larger tumulus. Patrick (1975:11) has suggested that the entire Brugh na Boinne complex (Newgrange, Knowth, Dowth, and all of their associated monuments) were the geographical focus for orientation of the many other passage graves in the surrounding region. Evidence for this is not, however, convincing.

Cursus monuments, associated with Stonehenge and several other henge enclosures, also may have astronomical orientations. They consist of very long parallel banks and ditches and so provide clearly defined alignments. Penny and Wood (1973) published the only detailed astronomical study of a cursus. They analyzed a set of interconnected monuments that extends for some 10 km (6.2 miles) on Cranbourne Chase in Dorset and found solstice and lunar standstill orientations.

Henge monuments, named for the low bank and ditch that surround the stones at Stonehenge, are earthen enclosures. Approximately 170 of these monuments have been identified in Great Britain and Ireland, and most of them do not enclose stone circles. In a few cases, however, excavation has shown that they once surrounded wooden post structures set out like the stone rings. The entrance through the enclosure at Stonehenge is roughly oriented on the midsummer solstice sunrise. An east-northeast orientation of the entrances of the majority of the surviving henges in Ireland therefore prompted me to consider an astronomical explanation. This work is still in progress, but at present it seems that the center of the entrance, as viewed from the center of the enclosure, would not have marked the significant astronomical events identified by Thom and others. It is possible that the orientation reflects only the direction of sunrise on the day the site was laid out, as was the case with some British churches until recent times. A similar explanation for generally southwesterly orientation of such monuments as the Irish wedge-shaped gallery graves has been offered.

Questions of Method

Interpretation of megalithic monuments requires us to ask exactly how they were used. We

Dominated by a monumental stone, carved with spirals and lozenges, the entrance to Newgrange also has a more subtle feature, a "window," above the door, that allows the winter solstice rising sun to shine into the inner chambers. The stone facing surrounding the entrance of the monument has been recently reconstructed and may differ significantly from its original appearance. (Robin Rector Krupp)

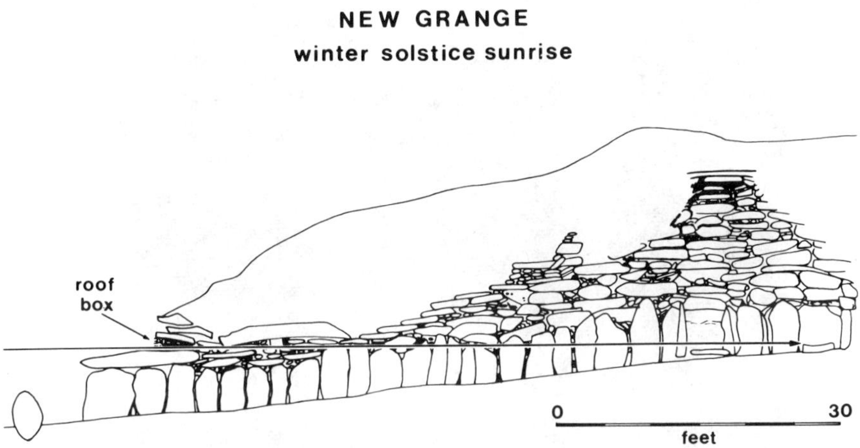

The upward slope of the passage floor prevents sunlight from reaching the innermost chamber of Newgrange, but the roof box above the door allows the winter solstice sun to penetrate into the heart of the monument. (Griffith Observatory)

The rising midwinter sun shines through the roof box over the entrance of Newgrange and illuminates the passage beyond. (Mike Bunn)

must first understand how the alignments were defined by those who set them out. Principles of observation may have varied from monument to monument or region to region. For passage graves the answer seems obvious. The event observed was illumination of the chamber, or perhaps some particular feature within it, by the sunlight shining down the passage. Lines of stones indicate the intended directions without ambiguity, but we don't know whether observers sighted along the left-hand side of the alignment, along the right, or across the top. These three choices can give different results. Finally which end served as the backsight, or did both?

For stone circles, cursuses, and henge earthworks, the possibilities become much wider. As Professor Thom (1967:93) has pointed out, so many possible alignments are available at many sites the choice of alignments included in an analysis is inevitably subjective. At Stonehenge observations made with the earlier components of the monument may have involved the views from the center and across the Heelstone and from the various corners of the Station Stone rectangle across the tops of the stones at the other corners. In Stonehenge's later period observations from the archways of the trilithons through corresponding gaps in the outer sarsen circle suggest a completely different approach. Thom (1967:93ff) discussed the various possibilities for other sites at some length. He assumed that the observer's position must be marked by a backsight, "a stone, a hole in a stone, a gap between two stones, or a pole at the center of a circle," and the foresight must be "a distant stone,...a pole at the centre of a circle...a distant mountain peak, a distant notch in the horizon, ...a rock far out at sea,...the fall of a steep island" (any of the prominent natural features being clearly indicated by a slab, an alignment, or an upright stone outlier at some distance from the backsight). It is only in those cases where distant horizon profiles are used that really precise measurements of solar and lunar motion can be made. Thom believes the locations of most stones in stone circles define the geometry of the ring rather than establish any astronomical alignments. A possible exception, the large recumbent slab in any of the so-called recumbent-stone circles of Aberdeenshire

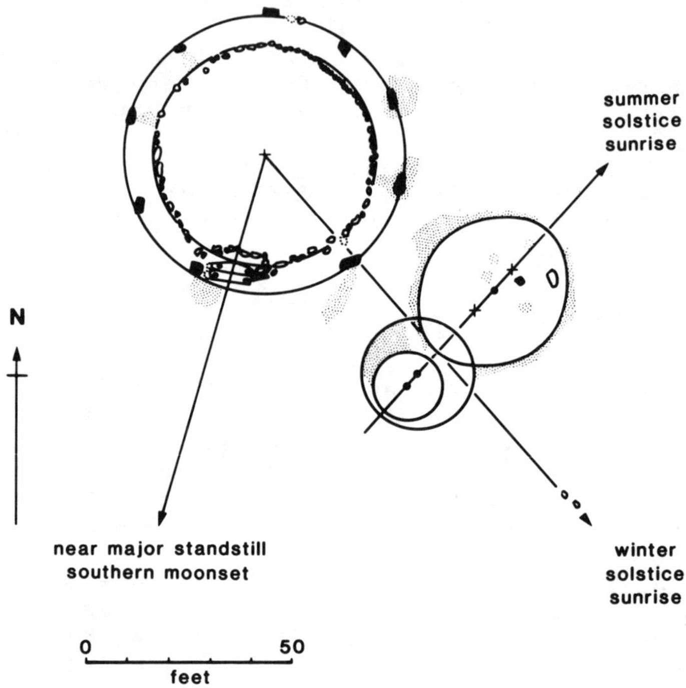

Loanhead of Daviot

A massive stone, lying on its side, distinguishes the recumbent stone circles of northwest Scotland. At Loanhead of Daviot, the recumbent stone is on the south-southwest arc of the ring. In Burl's interpretation, the stone here could have been a stage for the major standstill southernmost moon. (Griffith Observatory) and southwestern Ireland, may have been oriented with astronomy in mind. Although too large themselves to define an azimuth, these stones may have supported or helped locate less durable sighting devices (Thom 1967:56). Thom also suggests that diametrically opposed stones, rare in stone rings, possibly were intended as azimuth indicators. A critique of Thom's criteria and some suggestion for refinements were presented in 1977 in a report on a new survey of Callanish by J.A. Cooke and others. Additional criticism of Thom's approach was offered in 1978 in an article by Moir, and more recently Aubrey Burl(1980) has argued convincingly that the recumbent-stone circles of Scotland were oriented not to a rising or setting body but to the moon when it was high in the southern sky, likely for use at the maximum standstill when it would have appeared to skim along the top of the recumbent.

John Barber's (1973:26) investigation of the Irish recumbent-stone circles assumed that the relevant alignment is a monument's main axis. This line is defined by the center point of the recumbent slab and the point midway between the portal stones, a pair of standing stones opposite the recumbent. The portals create entrances to many such sites, and they usually are the tallest stones in such rings. Aubrey Burl (1976:176) assumed the same design for the Scottish rings.

Barber's study of the construction of the Irish recumbent-stone circles indicates the arrangement of stones in such monuments was obviously intended to be symmetrical. Positions vary from those expected, however, by ± 6 inches. This means that in the smaller circles the main axis can be defined only to a precision of $\pm 3^{\circ}$. Obviously, this limits the level of observational precision permitted at the site.

Penny and Wood, in their study of the Dorset cursus complex, assumed alignments from the centers of the cursus terminals to the centers of various long barrows and round barrows and lines between the barrows themselves. My own research of the Brugh na Boinne passage grave cemetery in Ireland implies a similar use of mound-to-mound alignments between the complex's three major passage graves.

Opposite entrances at the Dowth Hall henge, in Ireland, suggest an interest in the summer solstice sunrise. Here we are looking along the monument's entrance axis, toward the northeast. (Robin Rector Krupp)

Most of the earthen henge enclosures in Ireland do not incorporate stone circles, and I was first inclined to follow a procedure similar to Barber's (1973). I defined alignments in terms of the enclosures' centers and the midpoints of their entrances. In the few cases where excavations have been carried out and revealed postholes within the rather wide arc of an entrance, however, or where there are outliers evident, these additional features are not usually located on these centerline alignments. In addition, I have found myself asking whether one is intended to stand at the center and look out of the enclosure, or look from the entrance toward the center. For that matter, it may be that some horizon feature is intended from the center, or, in the case of enclosures with more than one entrance, the sightline may involve looking in one entrance and out another. Several of the enclosures have banks high enough to form an artificial horizon, and in at least one of these cases the entrance-to-entrance alignment seems to be relevant. It conforms to the summer solstice sunrise. This henge, at Dowth Hall, is especially interesting. A legend, recorded in early medieval times, tells us the sun was stopped in the sky to permit a monument to the gods to be built there.

Anyone who has worked with the megalithic monuments in the field probably will agree with Thom's statement that a certain amount of subjectivity is inevitable in the choice of alignments for analysis. Freeman and Elmore (1979) have offered, however, a test for the significance of astronomical alignments. It can be applied to the results of a survey of a *group* of sites for which alignments have been defined in a consistent way. Only three sets of data are currently adequate for testing, those of Thom, Barber, and Cooke et al. Freeman and Elmore do not feel significance is supported by their test. Where, I wonder, does that leave us? The correct answer is probably Heggie's (1977:453). He feels the orientations have a symbolic rather than observational value. In other words, while being roughly in the right direction for important risings and settings, or simply within the range of movement of the sun or moon, the alignments were never intended to be accurate enough for precise observations.

Other aspects of megalithic observational

Entrance orientations of Irish henge enclosures, plotted here in polar coordinates, do not seem to favor clearly any particular direction, although the northeast and southwest quadrants were more popular than the other two. Each henge is represented by its arc, and so differences in the radii in this plot are arbitrary and do not refer to the size of the henge. The length of each arc does indicate, however, the angular width of the entrance. Some sites have more than one entrance gap, but not all such gaps are proven to be part of the original designs. (Ronald Hicks)

techniques may have a bearing on the apparent inconsistencies and lack of exactitude. Hoyle (1977: 66) has argued that alignments on solar and lunar extremes at Stonehenge were deliberately offset to allow observation of the sun, or moon, as it first approaches and later retreats from the actual extreme position. This technique could permit an accurate determination of the correct date of the extreme position, the date midway between the dates on which the sighting line was crossed. This works better for the sun than for the moon, for the moon's day-to-day movement along the horizon is too quick.

We must also ask what part of the solar or lunar disc was observed: the upper limb, the lower limb, or the center? Hawkins (1965:111) argued for the lower limb. His Heelstone alignment required the sun's disc to be tangent to the horizon (or just grazing the top of the stone) on the summer solstice sunrise. Just about everyone else prefers the upper limb, or first gleam (last gleam for the setting disc), at least for the sun. According to Thom (1967:121) however, the moon's upper and lower limbs were both observed to reduce observational errors. In reply Moir (1979:125-126) has argued that it is very difficult to observe accurately either limb of the moon. The center of the disc, of both the sun and moon, usually is dismissed as difficult, if not impossible, to identify accurately without precise instruments unavailable to megalithic astronomers.

Of prime importance in any analysis of megalithic sites is the best technique for obtaining accurate surveys. Astronomical interpretation requires well surveyed plans, as does any study of the principles of a site's design and construction. In Megalithic Sites in Britain, Thom described his usual procedure. He uses "theodolite and tape, the orientation being determined by at least two time-azimuth observations of the sun" (1967:3). In overcast weather azimuths were determined by sighting on known distant points and by calculating orientations from their geographical coordinates. In a few cases, inaccessibility of sites ruled out the use of a theodolite, so a prismatic compass was used instead, in conjunction with a distant, known mark for orientation. These variations from the usual procedure necessarily limit the accuracy of some of the surveys.

Another difficulty develops in the measurement of horizon altitudes. Ideally these should be taken around the whole horizon, but in many cases essential horizon altitudes were not recorded initially and had to be estimated later from the Ordnance Survey contours. This, too, introduces errors. Thom recommends photographing the horizon, with the azimuth and altitude of two points in each picture being recorded to allow later measurements of the remainder with reasonable accuracy. If the azimuths indicated by the megalithic monuments are intended only as rough guides to the points of rising and setting of celestial objects, any small errors that may creep in using the methods described are of little concern. But if some of the sites were, as Thom suggests, intended for extremely precise observations, these small errors compound the problems described earlier. Cooke et al. (1977:117-118) estimate that the various inaccuracies introduced by limitations of the theodolite, uncertainty in the refraction correction, unfirm footing, parallax, and the like produced an error of roughly ± 1' in their declinations. Patrick (1979:80) mentions further errors that can be added by such things as growth of low vegetation in a nearby foresight. I might add, from my own experience in Ireland, trees have a habit of growing just where they can best blot out vital bits of horizon. This leads to roundabout surveying methods that multiply the chance of error.

Accurate mapping of the stones of a site involves other problems. Upon what positions should our calculations be based? Thom opts for the centers of the stones and for this reason carefully measures at or near the base of each (Thom 1967:4, 1971:122). He candidly admits, however, (e.g., Thom 1978), there are many problems in making even this determination. In some sites the stones are small and easily may have rolled out of place. In both Britain and Brittany, "The rows are ragged with gaps and places where the stones are close together but we do not know if this is due to stones having been removed...or if it represents the original state of the rows" (1978:256). Human interference certainly is a factor. Some of the stones have been reerected, not necessarily in their original positions. At Kermario "disturbance in nearly all the rows has been very great" (Thom &

Thom 1974:34). Smaller stones may have been shifted by repeated freezing and thawing of the soil through the centuries. Trees growing close to a stone can shift it (Thom & Thom 1978:3). Thom remarks, "No matter how carefully the surveys are made, accuracy in the distances is impossible," (1978:253), and adds, "No two people measuring the rows would get identical values: (1978:253). Thick vegetation sometimes makes accurate surveying extremely difficult. In commenting on the results from Carnac, Thom & Thom (1972:13) state, "The hope is that a sufficient number of the stones are near enough to their original positions to give, overall, a reliable statistical result. We shall see that this condition is met, but had the disturbance been much greater nothing could have been done. It is essential that no further error be introduced by even small inaccuracies in the surveys." Can their barely positive assessment be accepted?

In addition to the problems raised by the Thoms themselves, Aubrey Burl (1976b:9-10) points out a few more. Surface inspection can be misleading about the original number and spacing of stones. Some sites have had stones added in recent times and some entire monuments have been moved from their original position and reerected.

Megalithic Measurements

The difficulties in mapping bear directly on a second major aspect of Alexander Thom's work, his argument that the megalithic builders possessed standardized units of length: the megalithic inch (0.816 of a standard inch), a megalithic fathom (5.44 feet), a megalithic rod, a half yard, and most especially a megalithic yard of 2.720±0.003 feet (revised to 2.722 feet for England). In Megalithic Sites in Britain (1967) Thom provided a table of diameters for 145 stone circles and rings in support of the existence of this yard. He applied statistical procedures for determining the existence of a quantum, developed by Broadbent (1955, 1956). Thom found a unit and claimed further, "There must have been a headquarters from which standard rods were sent out" (1967:43).

I must confess that I do not believe in the standardized megalithic yard. In Thom's table the

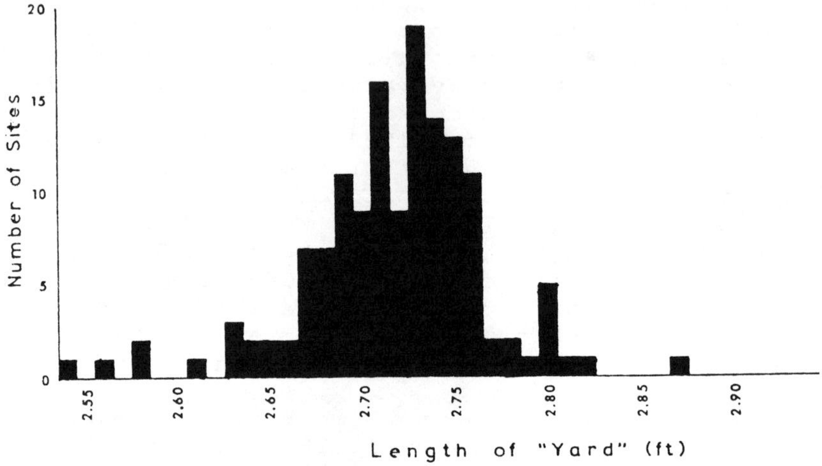

By assuming the stone rings were laid out in integral lengths that approximated Thom's megalithic yard but differed from it slightly from site to site and region to region, the measured dimensions of each ring provide a value for that local "megalithic yard." Analysis of A. Thom's (1967, pp. 36-39) tabulated diameters for 145 rings results in this distribution of values of the megalithic yard. There is a clear preponderance in the range 2.70 to 2.75 feet, but the total range is much wider. Rather than attributing this variation to errors in the original placement of stones and subsequent movement, we can ascribe this spread to local or individual variations in the standard of measurement. (Ronald Hicks)

diameters of the rings and the number of megalithic yards included in each are listed along with what he calls "residuals." These residuals are the amounts by which each diameter differs from an exact multiple. If these are taken into account, the actual range of values for the yard is considerable as can be seen in the histogram. The distribution is in fact entirely consistent with that one finds if a non-standard, traditional, nose-to-thumb yard is measured out by a large group of individuals. A table listing 178 measurements from stone rows (Thom 1978:254-255) reveals the same difficulty. In all the measurements listed in Thom's two tables, I can find only one where an exact multiple of the megalithic yard appears. In my opinion the megalithic yard exists only as a statistical artifact. In a complex statistical reanalysis of Thom's data, D.G. Kendall (1974:249) has concluded the existence of the standard fathom is supported--just barely--for the Scottish sites but it is definitely questionable in English and Welsh sites. Using similar methods Freeman (1976) and Burl & Freeman (1977) show that if regional groups are analyzed independently, a variety of lengths can be supported.

We must ask if Broadbent's procedures would make an imprecise, semi-standardized unit of measure look like a standardized one. It is obvious that some unit of measure must have been used in laying out the sites. Most likely it was one of the nonstandardized, traditional measures known to have been used throughout much of the world--the elbow-to-thumb cubit (half yard), the nose-to-thumb cloth yard, or the thumb-to-thumb fathom (Hicks 1977). Frank (1980) describes the use of just such a traditional system for laying out stone circles in recent times in Basque areas of Europe. Others have argued for pacing (Porteous 1973) or the height of one of the people involved in the construction (Hammerton 1971), but I feel the first of these would be too unwieldy since it would require repeatedly pacing out the same distance from the center or foci (also increasing the chance of error) and the second simply seems unlikely.

Megalithic Geometry

Thom's discussion in Megalithic Sites in Britain is undoubtedly the most comprehensive consi-

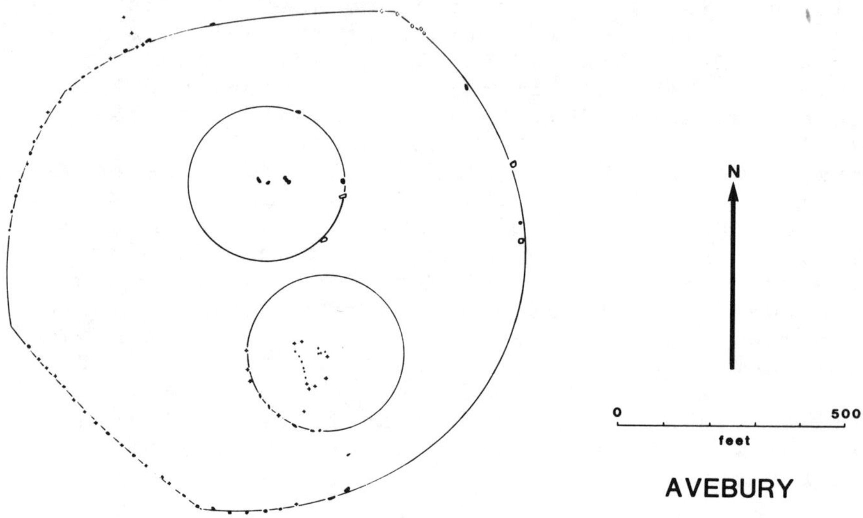

The Thoms explain the irregular shape of the great stone ring at Avebury as a composite design, based upon the megalithic units of measure and rules of geometric design they believe were in use in prehistoric Britain. Each segment of the ring is a circular arc that conforms to these rules and satisfies certain demands placed upon the entire plan. (Griffith Observatory)

deration of the geometry of the megalithic sites. In it he identified several shapes--circles, flattened circles (two types), egg-shaped rings (also two types), ellipses, and what he calls "compound rings." The variety of shapes, according to Thom (1967:27, 84), arose from the megalith builders' desire to base their constructions on Pythagorean triangles and to have as many dimensions of their monuments as possible, including circumferences, arranged to be integral multiples of the megalithic yard or, in the case of the compound rings, to have a circumference that was a multiple of 2½ megalithic yards.

To anyone who surveys megalithic sites, it quickly becomes obvious that the shapes listed by Thom do indeed exist. It also becomes obvious that the majority of sites deviate somewhat from being precisely circular, elliptical, or whatever. Skepticism about the megalithic yard implies an equal skepticism about designs involving multiples of this measure. Imprecision, which is evident in the shapes of the monuments, is most likely the result of the megalithic builders' inability to lay them out precisely to plan. This is most evident in the larger monuments. Of these, the most important example of a "compound" ring is the outer stone circle at Avebury, which is approximately 1100 feet in diameter. P.R. Freeman (1977), Aubrey Burl (1979: 179), and E.G.H. Crouch (1980) have also expressed their doubts about Thom's explanation of the irregular shape of Avebury's outer circle, and Burl cited archaeological evidence favoring construction of the circle *after* the surrounding bank and ditch. This would have made it impossible to lay out the monument as Thom (1967:89, 1978:30) suggests. Thom's theory requires that one arc be centered outside the earthwork. According to Burl, the prior existence of the two inner circles also would have greatly complicated the laying out of an accurate outer ring.

A more fruitful line of questioning concerns the development of the various shapes. Burl's summary (1976a:43) of the populations of the various shapes lists 600 circles, 150 flattened circles, 100 ellipses, and 50 eggs. This declining frequency corresponds to Cowan's (1969) conclusions on the probable order of development of the various shapes. Burl's (1976a:45-50) consideration of dating, based on architecture, radiocarbon dating, and associated artifacts, agrees. I.O. Angell (1976) has presented a satisfactory explanation for these shapes (which exclude the questionable compound rings) by demonstrating that all can be produced more simply than in Thom's theory by the use of from one to three stakes and a loop of line.

Megalithic Iconography

The "iconography" of the megalithic monuments is primarily what is expressed in associated petroglyphs, but a perspective from the air suggests a monument as a whole may be symbolic. For example, Avebury's two inner circles give the monument the appearance of a face. The Thoms have had little to say on this topic, beyond discussing the so-called cup-and-ring marks cut into many of the megaliths--depressions surrounded by concentric rings (Thom 1969; Thom & Thom 1978:45). They considered only the geometry and possible existence of a standardized unit of measure, in this case the megalithic inch. Since it involves only measurement, this certainly appears to be safer territory than the shaky land of interpretation of the symbolic intentions of a people 4000 years removed from us, although as the Thoms point out (1978:48), weathering and other damage have made even measurement difficult. Evan Hadingham (1974, 1975:136-151) has summarized our knowledge of such carvings to date, and Aubrey Burl has more recently discussed their possible astronomical implications (e.g., 1980:195).

The petroglyphs are found in their widest range in the passage graves, where motifs range from cup-and-ring marks to zigzags, spirals, penannular rings, lozenges, and a variety of less easily categorized forms. The motifs found in the Irish passage graves are discussed in detail by Claire O'Kelly (1973), but to date little has been done to try to interpret these. One form, however, has been called the "eye goddess" for years (Crawford 1957), though the original arguments for this no longer seem valid. Can we interpret the symbols astronomically (for example, in terms of their placement in the monuments)? Or can we assume sufficient continuity of tradition from the neolithic into early historic times, when recorded traditions give us some idea of the symbolism associated with certain deities, to venture a guess at their mean-

Archaeological evidence suggests that the massive earthen henge and deep trench at Avebury (seen here looking northwest, across the south and southwest arcs of the great ring, from the south bank of the henge) were built before the stones were set up on the central platform. If this be so, it is hard to see how the Thom rules of construction, which involve radii of curvature that extend far outside the banks, could have actually been followed. (E.C. Krupp)

Contrasting pillars and lozenge-shaped stones are paired in the massive megaliths of Avebury's Kennet Avenue. These shapes could be stylized genitalia, related in some way to rituals of fertility or renewal at prehistoric Avebury. (Robin Rector Krupp)

ing? The petroglyphs are fascinating. They certainly should be studied in more detail, but their interpretation is likely to remain highly speculative.

There is another aspect of iconography: the shapes of the stones themselves. Aside from some limited discussion of the stones in the Avenue at Avebury, little has been done. I strongly suspect that in many sites the stones' shapes are symbolic, though the effect of weathering through time has made the original shape of some stones uncertain. In the Avebury Avenue stones alternate, at least in some instances, along each side of and across the Avenue, from pillars to lozenge-shaped stones. Some think the monument and the shapes of the stones are linked to fertility ritual, with the lozenges representing female genitalia and the pillars male. But at other monuments other symbolism is clearly involved. At both Brogar and Stenness in the Orkneys one stone in the circle has what can only be described as a zigzag shape. The lozenge and zigzag occur, we know, in the petroglyphs. At the Orkney sites and at others, such as Drombeg in Ireland, the tops of many of the stones slope in different directions, very possibly to distinguish various sight lines from one another. The wide span of time covered by the passage grave art, cup-and-ring marks, and erection of the stones raises important questions about long-term continuity, but there can be no doubt that this is an important area for future study.

Finally, one other symbolic aspect of the monuments has been considered by Aubrey Burl (1976b): the numbers of stones. Although certain numbers of stones seem to have been preferred, these varied from region to region and had no connection with astronomical phenomena, megalithic units of length, or even the size of the monuments. Burl claims that the number of stones may have been related to nothing more than the counting system used (perhaps based on 4 or 6 in many cases) and to the methods used in construction.

Megalithic Society

Based on the archaeological evidence accumulated over the past century and a half, it has become rather generally accepted that society in neo-

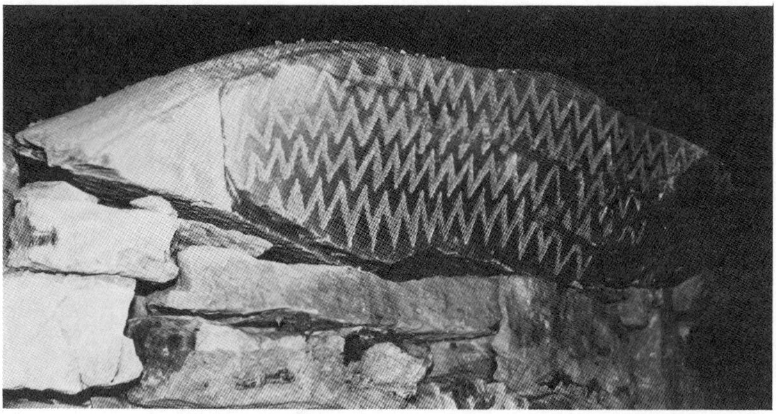

The zigzag is one of many symbols common in the prehistoric rock art of the British Isles. This flamboyant example is a lintel inside the Fourknocks I passage grave, about 17 miles north of Dublin. (E.C. Krupp)

Some prehistoric rock art appears to be celestially symbolic. This rayed disk is one of several carved on the face of a curbstone belonging to the Dowth passage grave, near Newgrange in Ireland. It is reminiscent of California Indian symbols illustrated elsewhere in this book, and although we cannot prove it represents the sun, it is reasonable to think it may. (E.C. Krupp)

lithic Britain was essentially pastoral, with scattered farmsteads rather than villages. For the Stonehenge area in Wessex the Early Bronze Age aristocracy, made wealthy by their control of the trade routes across southern Britain, was thought to have imported a Mycenaean architect to design their temple. Bristlecone pine recalibration of radiocarbon dating has now placed Stonehenge and the Wessex culture centuries earlier, and this picture of prehistoric Stonehenge is no longer feasible. Work by Geoffrey Wainwright (Wainwright & Longworth 1971) on a series of very large henge enclosures and work by Colin Renfrew (1973) and others on locational analysis and ethnographic analogy have provided the basis for new theories. We now ask whether megalithic peoples had a strictly tribal society, a chieftainship (which implies a degree of hereditary rule), or a theocracy dominated by an astronomically trained priesthood. The very large henges are now known to have been occupied, unlike their smaller counterparts. Were they monastic centers in which the priesthood lived and were trained? Something of the sort certainly seems to have existed in Iron Age Britain and Ireland two millennia later. The most elaborate development of these ideas is provided by Euan MacKie (1977) in his book Science and Society in Prehistoric Britain. The archaeological evidence (see, for example, Burl 1976a) indicates that a considerable amount of regional diversity existed and that different rules may have governed both architecture and culture in general in different areas. With this should be contrasted Alexander Thom's advocacy of standardized metrology and geometry under central control.

Summary and Conclusion

At this stage there are relatively few archaeologists who doubt at least that some megalithic monuments were used astronomically. On the other hand, there are also few who feel that this was their only use or that in fact all monuments of a particular type were used astronomically even to the same degree. In many, if not most, cases the alignments are likely to have been rough and symbolic, intended for ritual purposes rather than for precise observations. The question of precision dominates the problem of interpreting the megalithic monuments. How precise were they *intended* to be? The evidence is ambiguous at best, and many of the people working on these problems--myself among them--are not as confident that precision was intended as are the Thoms.

Finally, one very positive aspect of the study of megalithic sites is its interdisciplinary character. It pulls together physical scientists, behavioral scientists, and humanists, and forces them all to become familiar with one another's work. To truly understand these monuments we cannot look only at the astronomy, the geometry, the metrology, the iconography, or the archaeology and evidence for social organization. We must look at all of these. Those who work on any one aspect must be aware of, and make a serious attempt to understand, the work being done on all the others.

References

Angell, I.O.
1976 "Stone Circles: Megalithic Mathematics or Neolithic Nonsense?" Mathematical Gazette 60:189-193.

Barber, John
1973 "The Orientation of the Recumbent-stone Circles of the South-west of Ireland." Journal of the Kerry Archaeological and Historical Society 6:26-39.

Broadbent, S.R.
1955 "Quantum Hypotheses." Biometrika 42:45-57.

1957 "Examination of a Quantum Hypothesis Based on a Single Set of Data." Biometrika 42:32-44.

Burl, H. Aubrey W.
1976a The Stone Circles of the British Isles. New Haven: Yale University Press.

1976b "Intimations of Numeracy in the Neolithic and Bronze Age Societies of the British Isles (c. 3200-1200 B.C.)." Archaeological Journal 133:9-32.

1979 Prehistoric Avebury. New Haven: Yale University Press.

1980 "Science or Symbolism: Problems of Archaeoastronomy." Antiquity 54:191-200.

Burl, H.A.W., & P.R. Freeman
1977 "Local Units of Measurement in Prehistoric Britain." Antiquity 51:152-153.

Cooke, J.A., R.W. Few, J.G. Morgan, & C.L.N.Ruggles
1977 "Indicated Declinations at the Callanish Megalithic Sites." Journal for the History of Astronomy 8:113-133.

Cowan, Thaddeus M.
1969 "Megalithic Rings: Their Design Construction." Science 168:321-325.

Crawford, O.G.S.
1957 The Eye Goddess. London: Phoenix House.

Crouch, E.G.H.
1980 "The Avebury Circle: Its Geometry and Metrology." Science & Archaeology 22:32-34.

Duke, Edward
1846 The Druidical Temples of the County of Wiltshire. Salisbury.

Frank, Roslyn M.
1980 "Basque Stone Circles and Geometry. Archaeoastronomy (U.S.) 3:28-33.

Freeman, P.R.
1976 "A Bayesian Analysis of the Megalithic Yard." Journal of the Royal Statistics Society 139:20-55.

1977 "Thom's Survey of the Avebury Ring." Journal for the History of Astronomy 8:134-136.

Freeman, P.R., & W. Elmore
1979 "A Test for the Significance of Astronomical Alignments." Archaeoastronomy (England) 1:586-596.

Hadingham, Evan
1974 Ancient Carvings in Britain: a Mystery. London: Garnstone Press.

1975 Circles and Standing Stones. New York: Walker.

1981 "The Lunar Observatory Hypothesis at Carnac: a Reconsideration." Antiquity 55:35-42.

Hammerton, M.
1971 "The Megalithic Fathom: a Suggestion." Antiquity 45:302.

Hawkins, Gerald S.
1965 Stonehenge Decoded. Garden City, N.Y.: Doubleday.

1973 Beyond Stonehenge. New York: Harper & Row

Heggie, D.C.
1977 "Megalithic Astronomy: Fact or Fiction?" Quarterly Journal of the Royal Astronomical Society 18:450-458.

1972 "Megalithic Lunar Observatories: an Astronomer's View." Antiquity 46:43-48.

Hicks, Ronald
1977 "Thom's Megalithic Yard and Traditional Measurements." Irish Archaeological Research Forum. 4:1-7.

Hoyle, Fred
1966 "Speculations on Stonehenge." Antiquity 40:262-276.

1977 On Stonehenge. San Francisco: Freeman.

Kendall, D.G.
1974 "Hunting Quanta." In The Place of Astronomy in the Ancient World. F.R. Hodson, ed. Philosophical Transactions of the Royal Society of London, A, 276:231-266.

Lockyer, Sir J. Norman
1901 "An Attempt to Ascertain the Date of the Original Construction of Stonehenge from Its Orientation." Nature 65:55-57.

1909 Stonehenge and Other British Stone Monuments Astronomically Considered. London: Macmillan.

MacKie, Euan W.
1977 Science and Society in Prehistoric Britain. London: Elek.

Moir, Gordon
1978 "Standing Stones and Astronomy in the Outer Hebrides." Newsletter of the Northumberland Archaeological Group 2(1):13-23.

Moir, Gordon
1979 "Hoyle on Stonehenge." Antiquity 53:124-129.

Newall, R.S.
1959 Stonehenge, Wiltshire. London: Her Majesty's Stationery Office.

Newham, C.A.
1964 The Enigma of Stonehenge. Privately printed.

1972 The Astronomical Significance of Stonehenge Leeds: Blackburn.

O'Kelly, Claire
1973 "Passage-grave Art in the Boyne Valley, Ireland." Proceedings of the Prehistoric Society 39:354-382.

Patrick, Jon
1975 "Megalithic Exegesis: a Comment." Irish Archaeological Research Forum 2(2):9-14.

1976 "On the Interpretation of Carnac Menhirs: a Reply to A. & A. S. Thom's Defence." Irish Archaeological Research Forum 3(1): 33-35.

1979 "A Reassessment of the Lunar Observatory Hypothesis for the Kilmartin Stones." Archaeoastronomy (England) 1:78-85.

Penny, A., & J.E. Wood
1973 "The Dorset Cursus Complex--a Neolithic Astronomical Observatory?" The Archaeological Journal 130:44-76.

Porteous, Hugh L.
1973 "Megalithic Yard or Megalithic Myth?"

Journal for the History of Astronomy 4:22-24

Renfrew, Colin
1973 "Monuments, Mobilization and Social Organization in Prehistoric Wessex." In The Explanation of Culture Change: Models in Prehistory. Colin Renfrew, ed. London: Duckworth, pp. 539-558.

Smith, John
1771 Choir Gaur, the Grand Orrery of the Ancient Druids. Salisbury: Easton.

Somerville, H. Boyle
1912 "Astronomical Indications in the Megalithic Monuments at Callanish." Journal of the British Astronomical Association 22:83.

1922 "Instances of Orientations in Prehistoric Monuments of the British Isles." Archaeologia 73:193-224.

Stukeley, William
1740 Stonehenge: a Temple Restored to the British Druids. London.

Thom, Alexander
1967 Megalithic Sites in Britain. Oxford: Oxford University Press.

1969 "Geometry of Cup-and-ring Marks." Transactions of the Ancient Monuments Society 16:77-87.

1971 Megalithic Lunar Observatories. Oxford: Oxford University Press.

1978 "The Distances Between Stones in Rows." Journal of the Royal Statistical Society, A, 141:253-257.

Thom, A., & A.S. Thom
1971 "The Astronomical Significance of the Large Carnac Menhirs." Journal for the History of Astronomy 2:147-160.

1972 "The Carnac Alignments." Journal for the History of Astronomy 3:11-26.

1978 Megalithic Remains in Britain and Brittany. Oxford: Oxford University Press.

Thom, Alexander, Archibald S. Thom, & J.M. Gorrie
1976 "The Two Megalithic Lunar Observatories at Carnac." Journal for the History of Astronomy 7:11-26.

Wainwright, Geoffrey
1975 "Religion and Settlement in Wessex, 3000-1700 B.C." in Recent Work in Rural Archaeology. P.J. Fowler, ed. London: Butler & Tanner, pp. 57-71.

Wainwright, Geoffrey, & Ian H. Longworth
1971 Durrington Walls: Excavations 1966-1968. Society of Antiquaries of London Research Report 29.

John B. Carlson

5. The Nature of Mesoamerican Astronomy: A Look at the Native Texts

The prime use of Maya astronomy was to learn the habits of the celestial powers so as to make predictable the hazards of living under their influence--Floyd G. Lounsbury (1978:804)

The Pre-Columbian Mesoamericans, particularly the Classic Maya, were among the ancient world's great astronomers. As many popular accounts have asserted, these long-vanished civilizations formulated an astronomical calendar of unrivaled exactness and precision even by modern standards. The periodic actions of the celestial bodies were used to mark important dates in the Maya calendar, thereby fixing impending civil, agricultural, and religious events. The Maya preoccupation with the passage of time seems unparalleled in history.

A great deal has been written on this subject over the last century, but much of this literature lacks the alleged coherence and exactitude of Classic Maya astronomy. The fragmentary nature of the evidence and the inherent romance of "the great lost civilization in the jungle" provide reasons for this excess of speculative writing. However, during the last twenty years, several breakthroughs were made which give us clearer insights into the true character of native Mesoamerican astronomy. Of primary importance is the new ability to "read" Maya hieroglyphic texts in both the codex books and in the monumental inscriptions. Much more reliable source material has become available to scholars, and, due to the pioneering work of Proskouriakoff (1960) and others, we now know that the monumental inscriptions contain the recorded dynastic history of the Classic Maya ruling elite. A second major source of information is found in the new interdisciplinary study of archaeoastronomy, which investigates such astronomical evidence as is found in the remains of architecture and ancient city plans.

In this necessarily brief introduction to native Mesoamerican astronomy, we shall concentrate on the textual evidence with particular emphasis on the Classic Maya inscriptions and surviving Post-Classic codex books. Anthony Aveni, one of the pioneers in Mesoamerican archaeoastronomy, presents the latest discoveries in this field in a companion essay in the present volume. We shall begin with a very general overview of Mesoamerican cosmological thought and then focus more closely on the Maya, their numerology and calendrical astronomy. We shall explore such questions as: How good were the Mesoamerican astronomers at predicting celestial phenomena? What cultural purposes did astronomy (and astronomers) serve in society? And, how did Mesoamerican and Maya astronomy in particular function with the calendar in the total cosmological scheme?

Textual Sources

The inventory of genuine pre-conquest manuscripts now known is exceedingly small. Perhaps seventeen fragments of the painted, screenfold codex books exist, most being housed in the major library collections of Europe and the Americas. Various facsimile editions are available for study. These codices contain matters of ritual, religion, calendar, genealogy, history, mythology, astronomy, and cosmology--if we may be permitted to use these "Western" categories.

Among these codices are three, possibly four, Post-Classic Lowland-Maya pictorial manuscripts which contain the most detailed and revealing information about ancient Maya calendar and astronomy. Of the so-called Dresden, Madrid, Paris, and Grolier codices, the manuscript housed in the Dresden library is the single most valuable source. These books are not astronomical tables or ephemerides *per se*, but are ritual-calendrical almanacs used for divination and prophecy. Presumably they contain much that was copied from earlier books, including the primary mathematical and astronomical texts that must have been in use during the Classic period.

These Pre-Columbian sources are supplemented by over 400 native pictorial manuscripts of the colonial period and by the accounts of Spanish chroniclers. The best discussion of this great corpus of material is available in the surveys and catalogs of John Glass (1975) contained in the Handbook of Middle American Indians, Volume 14. The books of "Chilam Balam" from northern Yucatán and the Quiché Maya "Popol Vuh," or "Book of Counsel," are useful native sources in that they contain valuable mythological and cosmological accounts. With regard to the calendar, rituals, and hieroglyphic writing of the Post-Conquest Yucatec Maya, the sixteenth century chronicle of Bishop Diego de Landa, entitled Account of the Affairs of Yucatán, is the single most informative work.

The Classic Maya monumental inscriptions are the other extensive source of both textual and iconographic data. This material, found on carved stone *stelae*, altars and tablets, painted murals and funerary pottery, comprises a major source of information relevant to Mesoamerican numerology, calendrical astronomy, and cosmology.

Mesoamerican Cosmology: General Overview

The world-view of ancient Mesoamerica can best be characterized as an integrated conception of man, space, time, and deity in a stratified, hierarchical cosmos. Units of time, space, and the numbers themselves were personified or deified in a system seemingly in search of one grand, unifying number. Time was cyclical in structure, with greater periods built on more elemental cycles of numerological significance.

The order of each cosmos or creation was occasionally rent by a catastrophic destruction, a moment of chaos before the creation of a new order. The cosmogony of the new creation was one of emergence from the Underworld through a cave, a rebirth from skeletal, underworld existence. Sacrifice and the shedding of human blood to regenerate the day-to-night-to-new-day order of existence according to prescribed ritual was a vital part of this "interactive" cosmology.

The cyclical repetition of time and history can be seen in the prophecies that equally affected mankind and the cosmos. Man, represented by a god-like ruler, occupied a central position in this scheme. The monumental record shows that members of the ruling dynastic elite thought of themselves as part of the celestial order with a cosmic mandate providing the authority for the rule of their lineage.

Space

On the *horizontal* plane, Mesoamerican space was divided into four quarters with four prime world-directions. These directions were associated with colors, birds, trees, animals, numbers, deities, parts of the body, and units of time, all unified by various cabalistic schemes. The surface of the earth with its quadripartite structure was envisioned as the back of a great saurian earth-monster floating in a cosmic sea. The four-quartered structure of Tenochtitlán, the Aztec capital situated in Lake Texcoco, must have been a spectacular re-creation of this cosmological conception.

The diurnal and other cycles of celestial bodies overhead gave a sense of motion to the horizontal plane that was translated into a ritual circulation of colors and directions. This circulation, usually counterclockwise, was reflected in ceremonial processions and calendrically-related festivals, and a sense of turning was also deeply embedded in language and social structure.

In addition to the four (usually cardinal) directions, there was the Center, a fifth world-direction. The 5-direction cosmos, which is naturally extended in three dimensions to include the zenith and nadir, exists in many forms throughout the Americas.

In the *vertical* domain, the Mesoamerican world was a layered configuration with an underworld, surface, and heaven. Mankind's realm, the earth's surface, was the boundary between the diadic opposites of Upperworld and Underworld, a contrast which mirrored day and night, light and dark, and

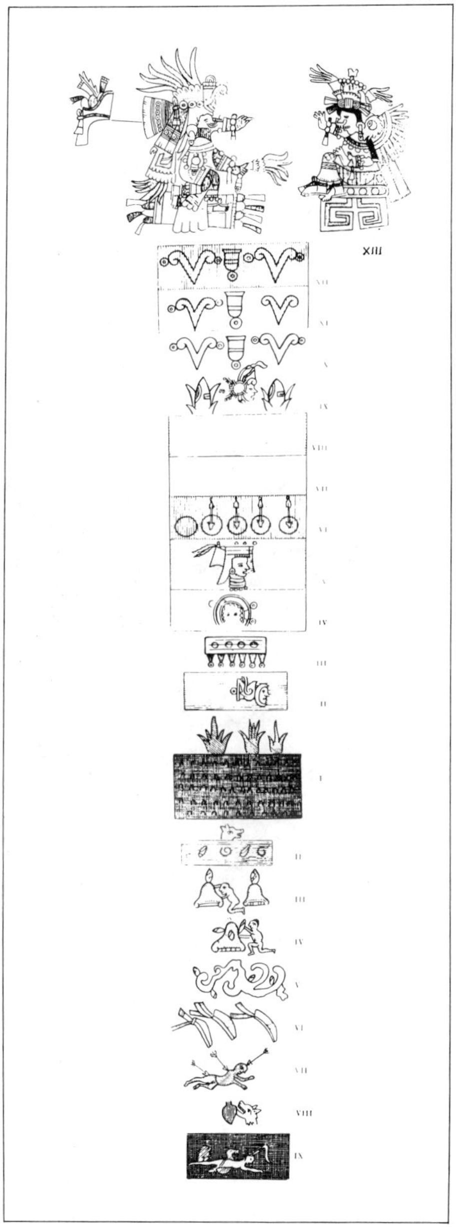

An Aztec depiction of the layers in a hierarchical cosmos places the earth (I) between 12 higher celestial realms and 8 lower underworlds. (from Codex Rios (Vaticanus A), after Nicholson, 1971, p. 407)

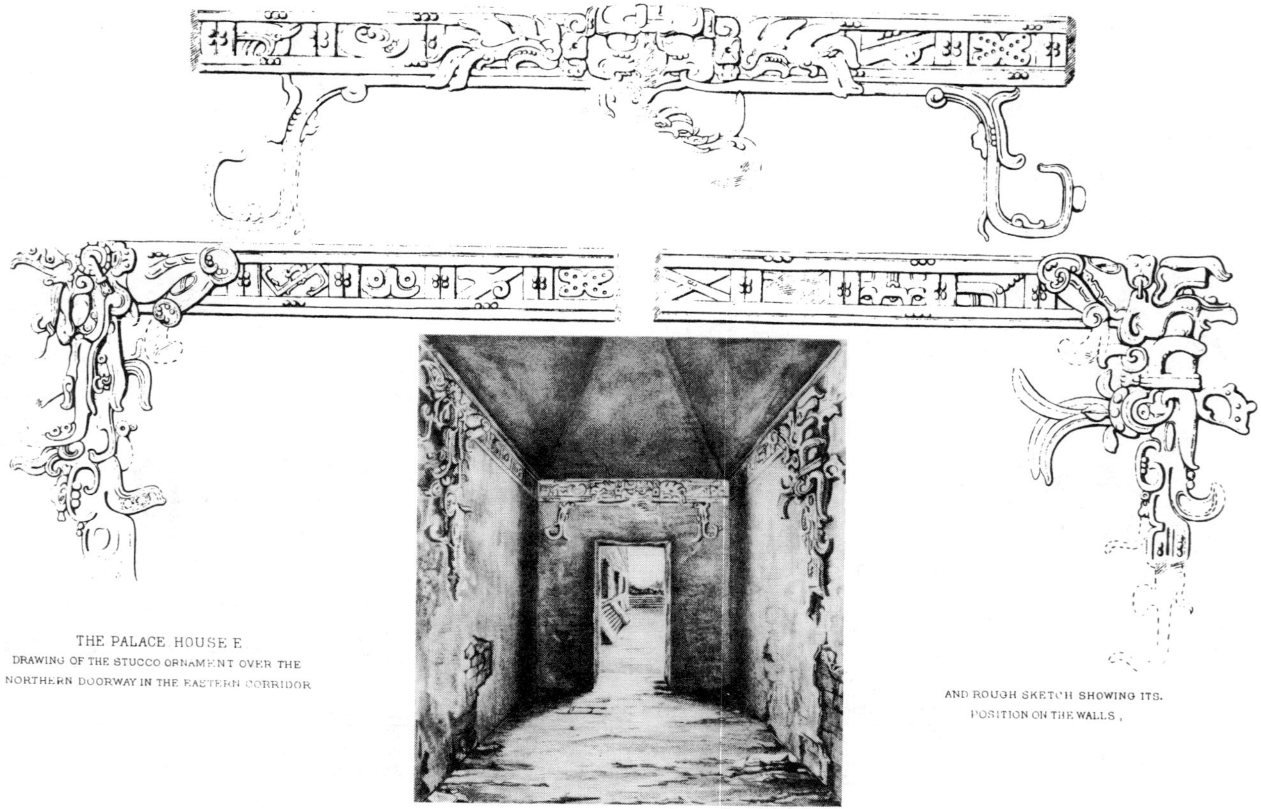

Ceiling molding in the form of a Maya bicephalic dragon with a skyband body circles around the eastern corridor of Palace House E at Palenque. The stucco ornamentation includes a companion creature, a composite "serpent bird," as well, perched over the northern doorway (after Maudslay, 1889-1902, IV, pl. 43)

life and death. To maintain the separation of these domains, the Maya conceived of four brother Bacabs, or World-bearers, who held up the sky at the four corners.

The vertical layering scheme had nine Underworld layers, each associated with a presiding "Lord of the Night", and thirteen Upperworld layers associated with thirteen "Lords of the Day." A version of this plan may be seen in an Aztec example from the pages of Post-Conquest Codex Rios. It has been suggested that this layering should be viewed as a pyramidal form: a seven-layer pyramid of heaven (thirteen steps up and down) and a five-layer pyramid of underworld (nine steps, down and up). The surface of the earth was labeled as number one in both upper-and underworld schemes, making people simultaneously part of both processions of deities.

The celestial bodies, deified entities travelling by day through the layers of heaven, were transformed into skeletal Underworld manifestations at night. As in the Quetzalcóatl legend, where the bones of Man from an earlier creation are revivified by the blood of the "Feathered Serpent" to form a new creation, the skeletal Underworld Sun also required blood sacrifice to rise anew at dawn. This life-and-death, flesh-and-bone duality is sympathetically linked to the astronomical transformation of day to night.

In the Central Mexican tradition, the original creator god was a male/female entity, Ometeotl, who resided in the thirteenth layer of Heaven, Omeyocán, the place of Duality. *Ome* is the number two in Nahuatl, the Aztec language, and *Omi* is the word for bone. Perhaps this play on words was intentionally understood to signify the blood-and-bone aspect of the dual creator of the gods and people.

Another cosmic pun relates the image of the serpent to the path followed by the celestial bodies through the sky. *Caan* or *Chan* is the word in Maya languages for both sky and serpent, so the sky is often symbolized by a two-headed dragon, a composite creature with serpent and bird characteristics. The body of this "bicephalic dragon" may be decorated with a series of symbolic elements, some of which relate to the sun, moon, and Venus. These bands of symbols have become known as "skybands" and, in many cases, they form a boundary or frame around scenes of cosmological importance.

To illustrate the Maya conception of space and related topics, we have the excellent example of the carved sarcophagus lid of the tomb of Lord Pacal, seventh century A.D. ruler of Palenque. Lord Pacal was buried in a specially prepared crypt within the magnificent nine-layered pyramid known as the Temple of the Inscriptions.

The central scene within the skyband frame may be viewed as space both in the horizontal and the vertical dimensions. The cross, which may represent the four world-directions, is actually composed of two crossed bicephalic dragons. At the navel of their cosmos, the Maya envisioned a massive "ceiba" tree with roots in the underworld and branches reaching to the highest heaven. Viewed in the vertical, the cross becomes this central world-tree, with a third more serpentine bicephalic dragon draped over its branches.

Centrally placed in the composition, Lord Pacal is seen seated on the west-facing monster head of the two-headed dragon. The upturned head faces east (the prime Mesoamerican direction) with the composite "serpent bird" embodying cloud and sky perched on top. The massive head below is a solar deity with "Kin" sun symbol in its forehead. It is seen with fleshed eyeballs and nose, but the lower portion of the jaw is skeletal, probably signifying its entrance into the underworld. In the same scene, Pacal, riding on the head of this solar monster is entering the open jaws of the earth monster along with the dying sun at sunset. Linda Schele (1977) has suggested that the death of the lord is being symbolically represented by sunset in the diurnal cycle, and by winter solstice sunset in the yearly cycle. The entire depiction represents the death and transfiguration of the lord; the world-tree rising behind Lord Pacal is a symbol of resurrection, and it contains the face of the sun god emerging into light and life after its skeletal underworld sojourn.

In the spatial composition of the sarcophagus lid relief, Lord Pacal himself embodies the comple-

An elaborately carved sarcophagus lid was discovered in the burial chamber within the Temple of the Inscriptions at Palenque. Many of the symbolic elements have cosmological and astronomical meaning. (after A. Villagra in Ruz 1970, fig. 74)

The first page of the Codex Fejérváry-Mayer diagrams the four quarters of the world with their associated colors, plants, animals, and Lords of the Night. The 260-day almanac is represented by the dots on the perimeter as well as by pertinent day-name glyphs.

tion of a generation in his dynastic lineage, and also symbolizes more general cycles of Maya time and space. His head is balanced symmetrically on the left side of the cross by the Maya completion sign--usually used as the symbol of completion of the 260-day divinatory almanac. This completion symbol may be compared with the space-time diagram from page 1 of Central Mexican Codex Fejérváry-Mayer. Here, as will be discussed later in more detail, the twenty groups of thirteen days of the 260-day calendar are wrapped around the four-quartered space completing the sacred almanac.

In summary, the scene shows the death and rebirth of Man within the cyclical continuity of the cosmological world-order and the dynastic lineage. The sarcophagus lid relief, like the Fejérváry-Mayer diagram, is a unifying mandala of the Mesoamerican universe, with spatial transformations eloquently linked to those of time.

Time and the Calendar

The day was the fundamental unit of Mesoamerican timekeeping. The Classic Maya developed the most elaborate of the Mesoamerican calendar schemes and it was based on the day, called a "kin", and larger numerologically ordained units of time. The harmonious commensuration of ever-greater units of time seemed to be the goal of Maya calendar building, and the largest cycles achieved truly cosmological scope. Maya calendar-priests are known to have made calculations amounting to hundreds of millions of years, accurate to the day.

The numerology within Maya calendar and computation reflects their view of both humanity and the *cosmos*. It is based on the perceived geometry (and periodicities) of *space*, the *human form*, and various *astronomical* (synodic) *periods*. This was a distinctly anthropomorphic scheme; Man was literally the measure of all things.

Counting was done in an arithmetic system of base twenty--a "vigesimal" system--with a base five subsystem. The Maya required only three symbols: a "dot" for one, a "bar" for five, and a shell symbol for zero or place-filler. Because the Maya employed place notation and developed the concept of zero, their system was perhaps the first advanced arithmetic, equal in quality to our own.

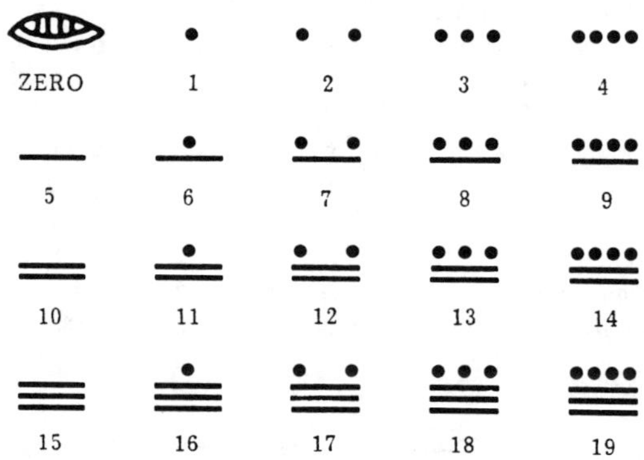

The Maya bar-and-dot counting system is based on the number 20. These examples show how bars (5), dots (1), and shells (0) are combined to generate the numbers 0 through 19.

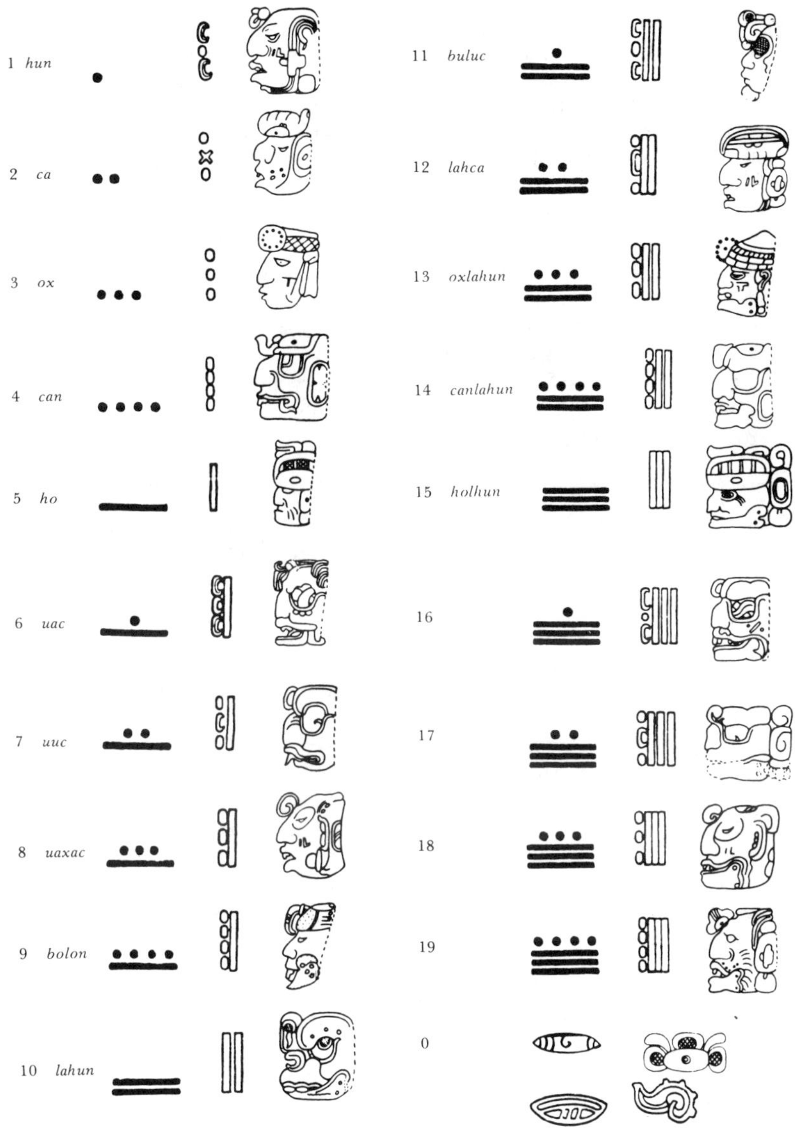

The Maya represented numbers with more elaborate symbols, too. Tabulated here with the simple bar-and-dot forms are an alternate version of the bar-and-dot scheme and the set of head-variant number symbols. The names of the numbers are from the Yucatec language. (after Kelley, 1976, fig. 1)

Maya numbers were also represented in anthropomorphic form. Each number from one to thirteen was actually a personage. For example, the Sun God could represent the number four and the skull of the Death God represented the number ten. These thirteen personified numbers may well be the "Oxlahun-ti-ku", the thirteen Gods of the Day and Upperworld.

The numbers above thirteen, and sometimes thirteen itself, were shown in composite form just as our "fifteen" means literally five and ten. In this case, the Maya would take the personage of the number five and give him the Death God's skeletal lower jaw to create the number fifteen. This composite number series may possibly represent a sequence of Underworld deities.

These personified numbers were also represented on the monumental carvings as abbreviated "head" forms or complete "full-figure" glyphs. The full-figure forms are used in calendric contexts where the god of the number actually bears a unit of time on his back. A number-god would carry the burden of time (itself usually personified) until the completion of that cycle of time and then pass it along to the next number-god to carry. The passage of time was thus perceived as a cycle of burdens borne by a repeated sequence of number-gods.

The vigesimal system is certainly based on the geometry of the human body with its four extremities each having five digits. The Yucatec name for the fundamental "month" of twenty days was *uinal*, which derives from the word for man, *uinic*. *Uinal* is also related to the Maya word for the moon, so it was used in the same sense as our word "month." Indeed, one form of the Maya glyph for both twenty and uinal is the moon symbol.

Before investigating the modifications of the vigesimal counting scheme when applied to calendrial purposes, we must examine the numerologically important factors up to and including the "completion"of twenty, namely 2, 4, 5, 7, 9, 13 and 20.

Two is of course the duality of the sexes, the bisymmetry of the human form, day and night, life and death, the "yang and yin" of this world-view. A "binary" cycle is the most basic.

This full figure glyph on Stela D at Copán is a "sun god" Lord of the Underworld who carries a bundle by tumpline in this Maya long count inscription. (E.C. Krupp)

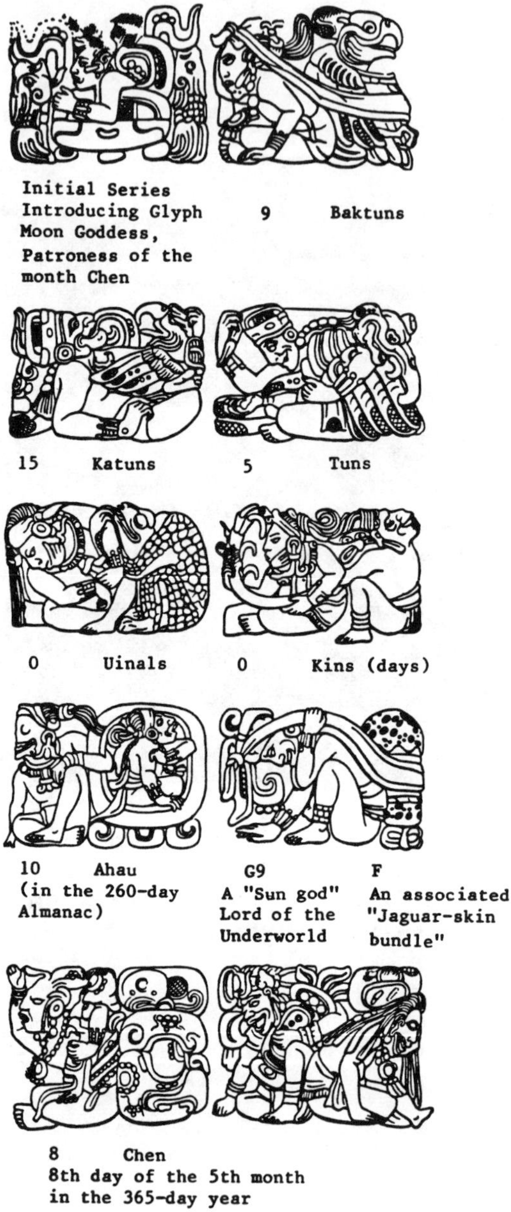

These figures are actually calendrical glyphs. They stand for certain numbers of various Maya Time intervals: Baktun (144,000 days), Katun (7200 days), Tun (360 days), Uinal (20 days), and Kin (1 day). This inscription is on Stela D at Copán, Honduras. (after Kelley, 1976, fig. 3)

There are *four* quarters of the body and the cosmos. Rather than the four cardinal directions, some Mesoamerican sources suggest the four solstice rise and set directions as defining the boundaries of the four quarters. With the thumb opposed, the four digits of the hand count out the quarters of the Mesoamerican mandala.

The importance of *five* derives from the hand and from its function as one quarter of twenty. Spatially, five is the fifth direction or center, the navel of the earth.

There are *seven* prime directions in the three-dimensional Mesoamerican conception--the four cardinal directions plus center, zenith and nadir. Seven is also the number of "classical planets" or moving lights in the sky that give their names and number to the seven days of our week of cosmological creation. It has also been suggested that the Mesoamerican "earth" had seven layers.

Nine is the number of Lords of the Underworld and Darkness. As mentioned previously, the nine-day sequence may have a basis in the number of steps up and down a five-layer pyramid. It may also relate to a "cubic" structure of four upper-world quarters and four Underworld quarters with Man at the center.

The Mesoamericans venerated *thirteen* Lords of the Upperworld and Day. This all-important magic number may possibly have a spatial interpretation in the layered, cubic structure consisting of four upperworld quarters, four earth quarters, four underworld quarters and the center. The cycle of *twenty* has already been described.

The greater calendric periods were all composed of these numerological factors. The 260-day almanac, 365-day "vague year," 52-year "calendar round" and the 13-Baktun "long count" were particularly important periods found in the Classic Maya monumental inscriptions and will be described here in some detail.

The *260-day calendar*--often called the sacred or divinatory almanac--was the most ancient and fundamental composite cycle known in Mesoamerica. It is still in use today in several Maya areas of

Mexico and Guatemala. One defining characteristic of the culture area of Mesoamerica is that it encompasses those peoples who used or were influenced by the 260-day almanac. Individuals were named for the day on which they were born in this cycle. Their fate and fortune were linked to the astrological properties associated with the day. Why should this seemingly odd period of time have gained such an important position?

The 260-day period is derived from the commensuration of the cycle of the thirteen deified numbers with the month of twenty named days. Though the ancient Mesoamericans would hardly have viewed this in a mechanistic way, we may constructively picture two intermeshing gear wheels, one with teeth numbered one to thirteen, the other with the twenty named days of the uinal. Any given combination of *number* and *day* repeats after 260 clicks of the gears.

The sacred almanac also represented the completion of the four-quarters structure of space. The world-diagram of Codex Fejervary-Mayer shows the twenty groups of thirteen dots each cycling around the four quarters with associated colors, birds, trees and deities. This shape has inspired the Mesoamerican symbol for cyclical completion and zero, as can be seen in the calendrical inscription on the east side of Stela E at Quirigua.

Though astronomical motivations have been suggested, one of the most convincing reasons offered for the overriding importance and universality of the 260-day sacred almanac lies in its function as a "canonical conception horoscope." 260 days is the approximate length of human gestation. At least in a ritual sense, the day of conception would have had the same auguries as the day of birth. One's name day in this system fixed the individual in this cabalistic scheme of generation and birth.

The second reason for the prominence of the 260-day cycle is astronomical rather than human. Quite fortuitously, it is neatly commensurate with several important celestial periods. As we will discuss later, the *double almanac* (2x260=520 days) just happens to closely approximate three lunar half-years (3x173.33=520 days). That the

Maya noticed that "eclipse possible" periods came only at three evenly-spaced times in this double almanac is evident in the structure of the Dresden Codex eclipse tables.

Furthermore, the *triple almanac* of 780 days corresponds closely, and again most fortuitously, to the synodic period of Mars: 779.94 days. There are triple almanac tables in the Dresden Codex that have been speculatively associated with Mars, and though this association is problematical, the Maya can hardly have failed to notice and appreciate this numerologically pleasing coincidence.

Probably for these reasons, reinforced with additional commensurations, the 260-day sacred almanac gained its position as the supreme calendrical cycle of social regulation in the Mesoamerican world.

The Mesoamerican *365-day year* was composed of 18 months of 20 days each, with an added "epagomenal" month of five days. This "vague year", as it is often called, fell short of the true tropical (solar) year by 0.2422 days. Thus New Year's Day would drift backward relative to the tropical year until it would again resynchronize after 1508 years, a fact apparently well understood by the Maya.

Mesoamerican peoples were interested in the commensuration of the 260-day Almanac with the 365-day year. Any given configuration will repeat only after 18,980 days, or 52 years of 365 days each. This 52-year cycle or "calendar round" was a cosmological cycle of major proportions. It was thought that the world could come to a violent end at the completion of a calendar round. Series of rituals and sacrifices were performed at the approach of the day to ward off cosmic catastrophe. One of these was the Aztec New Fire Ceremony, held atop the "Hill of the Star" near Mexico City. Timed by the passage of the Pleiades through the meridian, a victim, stretched to the four quarters, was sacrificed by removing his heart. A new fire was kindled with a "fire drill" in the chest cavity and distributed by runners throughout the Aztec empire. It has also been suggested that the practice of rebuilding temples in Mesoamerica was generally timed to this 52-year period.

The greatest calendric cycle generally used in Mesoamerica is called the *"long count"* because it specifies a period of over 5000 years. The vigesimal unit of this count is another "year" of 360 days which the Maya called the "Tun". This Tun is composed of 18 20-day Uinals. Twenty Tuns (7200 days) was called a "Katun" and twenty Katuns is a "Baktun" of 144,000 days. Although there were higher multiples of the Tun known, such as the "Alautun" (160,000 Baktuns, or more than six million years), a cycle of 13 Baktuns seems to have been the greatest period used in standard Maya inscriptions.

We cannot with absolute certainty correlate the Classic Maya calendar with our own. But if the so-called Goodman-Marinez-Thompson correlation is correct, the previous cycle of 13 Baktuns ended in the year 3114 B.C., and the current series began. The Classic Maya lived well after the completion of 10 Baktuns. We are still in this 13 Baktun era; it will end in our year 2011 A.D. near Christmas Day. Although it is certain that the starting day (called 4 Ahau 8 Cumhu in the calendar round) of the long count in 3114 B.C. antedated the development of Maya civilization by over 3000 years, its creation epitomizes the grand scope of the Pre-Columbian conception of time.

Classic Maya Monumental Inscriptions

A typical Classic Maya monumental text begins first with an elaborate calendrical statement. This may take up a whole side of a stela or carved stone monument. The calendrical passage is followed by further inscriptions that are now known to concern the events, rites, and mythology of dynastic ritual.

An example of such a lengthy calendrical text may be seen on Stela E at Quirigua, Guatemala. The glyphs are read from left to right in pairs and from top to bottom. This date begins with the oversized "Initial Series Introducing Glyph". The head in the center of this glyph is the patron of the month of Cumhu in the 365-day year. This is followed by the five glyph blocks in position A3-A5 which tell us that the day is 9 Baktuns [A3], 17 Katuns [B3], 0 Tuns [A4], 0 Uinals [B4] and 0 Kins (or days) [A5] since the beginning of the cycle. Note that the periods of time are in their head-variant forms, but the coefficients are in the

Stela 3 at Tikal is carved with a typical long count inscription. These commemorative stones seem to mark important dates in the dynastic histories of the Maya. (Robin Rector Krupp)

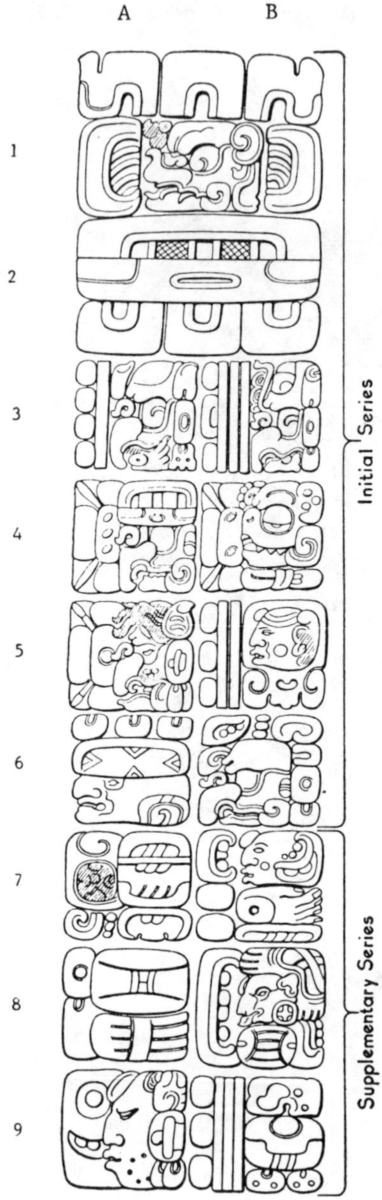

Maya long count inscriptions are, in large measure, sequences of calendrical glyphs that sum to the total number of days elapsed since the starting date for the entire system to the date of the inscription. Other information, some of it pertaining to the moon, may also be included. This calendrical text comes from the east side of Stela E Quirigua, Guatemala. (after Morley, 1961, fig. 25)

simpler bar and dot (and completion sign) notation. Such a date is very much like our 1980 A.D. This is now followed [at B5] by a date 13 Ahau, the last day of the 260-day Almanac.

The next face [at A6] is that of the ninth god in the cycle of Nine Underworld Lords, followed [at B6] by an associated glyph. The next five blocks [A7-A9] relate to the Maya lunar calendar and are called Glyphs E and D, C, X, B, and A, respectively. The elaborate and accurate Maya lunar calendar used 29-or 30-day months and placed them in groupings of six months called lunar half-years, which are related to an eclipse prediction scheme. In this inscription, these glyphs tell us that the day in question was a new moon (Glyphs E and D, [A7]), in a 29-day month (Glyph A, [A9]) which was the second month (Glyph C, [B7]) in the current lunar half-year. The last glyph block [at B9] describes the date as the 18th day of the month Cumhu in the 365-day year, which agrees with the Initial Series Introducing Glyph that previously named the patron of Cumhu.

As if this were not enough, these calendric preambles sometimes contain a further calendrical statement, the position in an 819-day cycle. This was actually a 4x819 or 3276-day cycle, since each station in the cycle of four was associated with the four world-directions and colors in a clockwise pattern. Why use an additional 819-day calendar? 819 is factored by the "magic" numbers 7x9x13, all important elemental cycles related to the layering of the vertical cosmos. This esoteric calendar was only employed by the elite to further specify events of dynastic concern.

The grand, quasi-astronomical Maya calendar described above was perhaps the most elaborate frame of time ever created by any civilization. The essence of a Classic Maya text is, however, historical. The inscriptions that follow the date speak of births, deaths, heir designations, accessions to rulership, battles, captures, sacrifice, and anniversaries of such events. The lords and their relations, both immediate kin and mythical ancestors, are described by elaborate titles and their relationships are often linked by calendrical numerology. Cosmic importance and auspiciousness was given to an occurrence by carefully controlling the numerological relationships. The calendar-priest/astrologers must have calculated the propitious days for marriages and military forays. It is also quite likely that they were able to manipulate history after the fact, to invent favorable days for events such as births and deaths which are usually beyond human control. Astronomy served this essential purpose in elite as well as lower-class Mesoamerican society. The astronomers, who were probably both calendar priests and astrologers watched the regularities of the celestial movements and also interpreted the omens, including transient phenomena such as comets, meteors, and eclipses. The Maya astronomer probably was able to predict eclipses, or at least knew when they were likely. This power to predict must have given these persons immense power over the ruling elite and the common man alike.

Floyd Lounsbury (1976) has been able to show that calendrical numerology plays a dominant role in some of the more extensive mythological and historical texts at Palenque. For example, he has demonstrated that there are numerologically contrived "distance numbers" that relate the dates of historical events in the life of a named individual to astrologically similar occurrences for mythological ancestors in a previous cosmological era. This would effectively create a cosmic charter or mandate for the rule of the dynast and his royal lineage, a form of "divine right of kings". It is why there are lengthy calendrical preambles placed before the "historical" texts; they serve to establish a cosmic frame of time for the arena of dynastic events.

Classic Maya Codex Books

How good was Classic Maya astronomy? What types of celestial events could be predicted and how precisely? How closely was the quasi-astronomical calendar related to true ephemerides? How did this astronomy develop and where was it going?

For the best clues to the answers to these questions, we must look at the surviving Post-Classic Codex books, in particular the Dresden Codex. We will briefly examine two crucial sections of this key divinatory almanac: (1) the eclipse tables (and lunar calendar) and (2) the ritual

Venus calendar. Both of these pertain to calendrical numerology as much as they do to real astronomy.

The Dresden Codex is not a primary astronomical text but an almanac devoted to ritual, divination, and prophecy. It dates from approximately the twelfth century A.D. and was composed in Northern Yucatán, perhaps at the great center of Chichén Itzá. It clearly shows evidence of having been painted over and rewritten at least once.

The Dresden Codex contains many 260-day almanac tables relating to such aspects of Maya life as rain-making, agriculture, fertility and calamity. The 260-day almanac is also fundamental to the most elaborate Maya astronomical tables known--the eclipse tables and the Venus calendar--and an understanding of this is crucial to our appreciation of the Maya treatment of time. The internal structure of these tables strongly implies that there once must have existed actual astronomical texts in the Classic period from which these divinatory manuals were composed.

The Maya Dresden Eclipse Tables

Pages 51 to 58 contain the eclipse reckoning tables of the Dresden Codex. An analysis of the divisions of the table shows that it contains 177-day (occasionally 178-day) and 148-day intervals. This identifies it as a lunar ephemeris containing groups of six and five lunations repectively. (The lunation, or synodic lunar period of phases, is more exactly 29.53 days.) The distribution and total number of six and five-month periods show that the lunar tables are organized for the prediction of times of potential solar eclipses. Nine five-month groupings are distributed among sixty of the six-month groupings in such a way that there are three major subdivisions of 135 lunations each, or 405 total lunar months, in the cycle. This is 11,960 days or five times the "Palenque moon ratio" in which 81 lunations equal 2392 days. This ratio gives a value for the lunar month of 29.530864 days, remarkably close to the correct value of 29.530588 days.

Eclipse prediction, whether solar or lunar, is based on the knowledge of the times of syzygy of the sun, moon, and earth. In some form, one needs to know the average length of the lunar month (29.530588 days), the average interval of time between lunar node passages (173.30906 days), and the so-called "ecliptic limits" for solar or lunar eclipses. For a solar eclipse to occur, the phase of the moon must be nearly full, and the sun and moon must be near either the ascending or descending lunar node, the points where the moon's orbit crosses the plane of the earth's orbit around the sun. In this case, the limits are from 15 to 18 days around the time of node passage. As we shall see, the numbers derived by the Maya were surprisingly close to the modern values.

One would expect that the eclipse prediction system would be based on the knowledge of the commensuration of the lunar month to the solar (tropical) year so that one could relate the cycle of possible eclipses to the seasons of the year. Indeed, the ancient Babylonians knew of two important such relationships, the "Saros" and "Metonic cycle" whose special significance seems to have escaped the Maya. The Saros relates 38 eclipse half-years to 223 lunations. This 18 year and 11 day period shows the relationship of 19 revolutions of the Sun with respect to a node to 223 lunar months. The so-called Metonic cycle relates 19 solar years to approximately 235 lunar months. Though these are important ways that we in the Western tradition might look at eclipse cycles, the Maya based their system not on the solar year but on the all-important 260-day almanac. As previously mentioned, they apparently observed over a period of time that quite fortuitously, solar or lunar eclipses only occurred within three symmetrically-placed sectors of their 260-day calendar. This means that three eclipse warning intervals would fall within the double almanac of 2x260=520 days. This effectively gives a Maya eclipse half-year period of 520/3 or 173.33, which is a fine approximation to the correct lunar node passage interval recorded above. The overriding importance of the almanac apparently prevented the Maya from penetrating more deeply into the astronomy of eclipse prediction. They were more concerned with the numerological commensurations within their calendar than with the geometrical and mathematical relationships of positional astronomy.

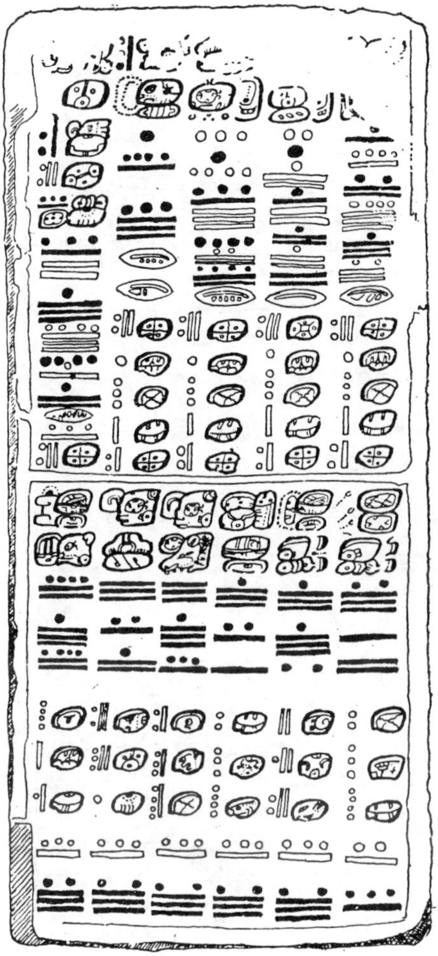

Dresden Codex page 51

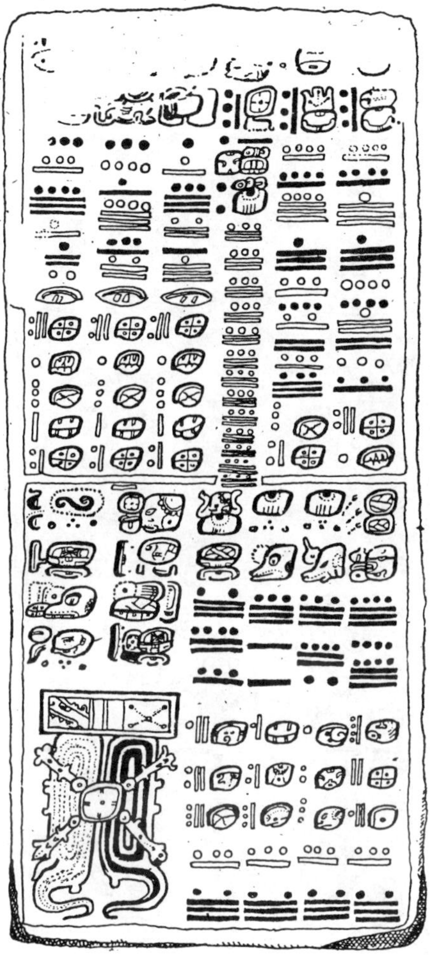

Dresden Codex page 52

Eight pages of the Maya Dresden Codex are numerical tables related to the recurrence of eclipses. The tables are organized into 177-day and 148-day intervals which correspond 6-lunation and 5-lunation periods that are useful for this purpose. The Maya estimate for the length of one lunation is based upon the equivalence between 81 lunations and 2392 days, and the Dresden Codex eclipse tables cover a span five times this long: 11,960 days = 405 lunations. (after Villacorta and Villacorta, 1976)

Dresden Codex page 53

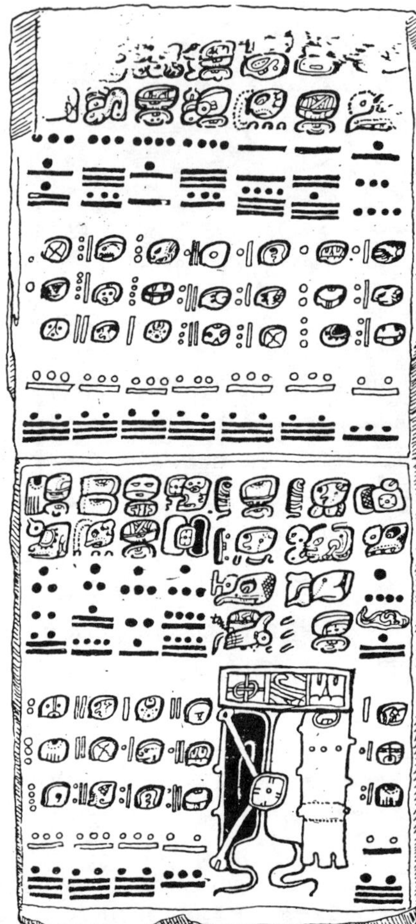

Dresden Codex page 54

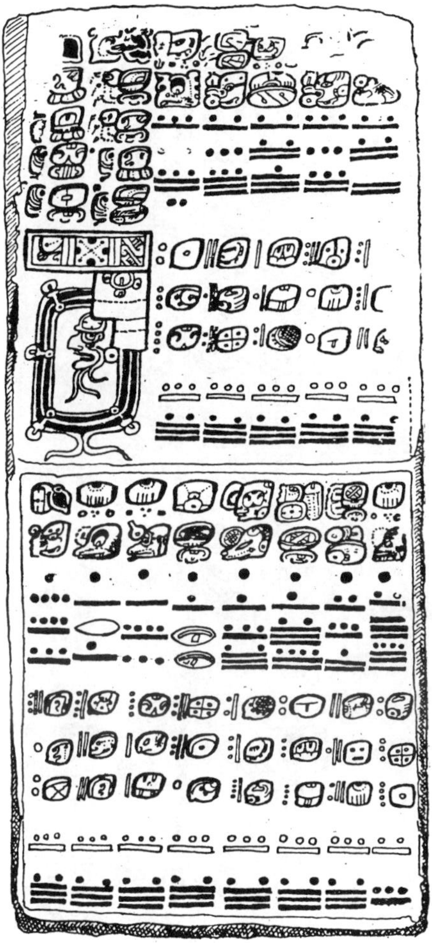

Dresden Codex page 55

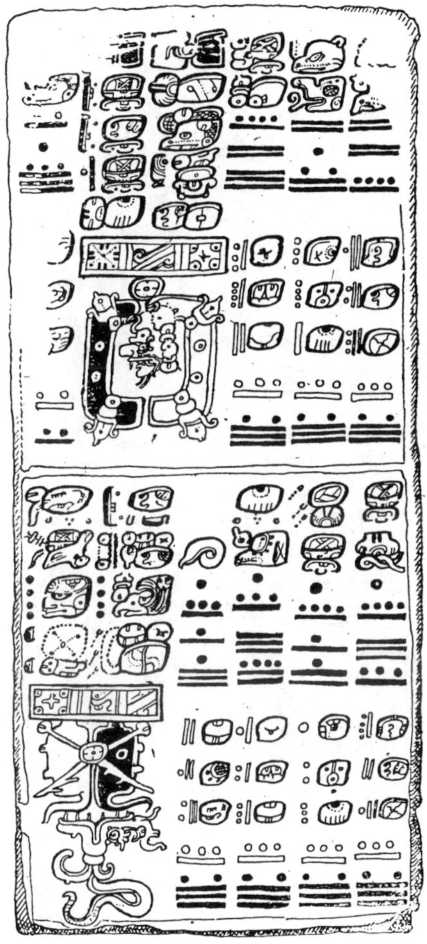

Dresden Codex page 56

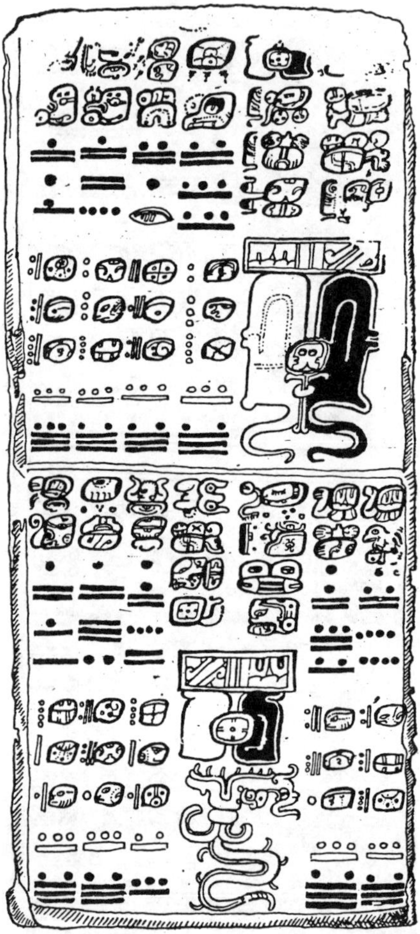

Dresden Codex page 57

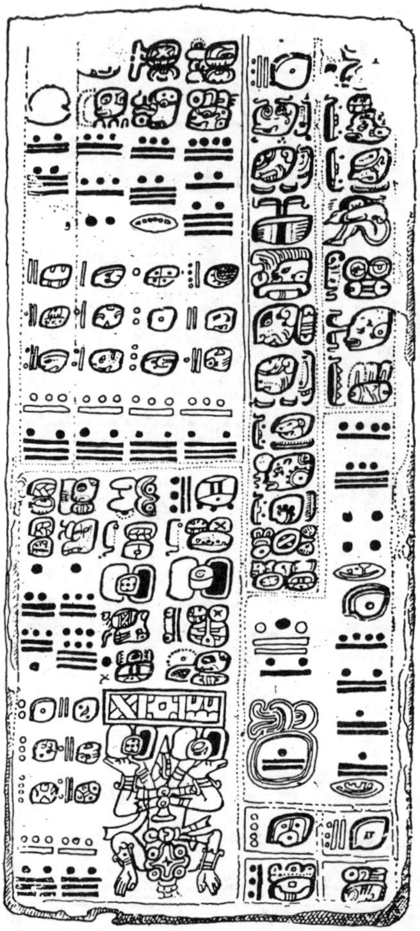

Dresden Codex page 58

Of course, any such scheme of exact numerical ratios of whole days is always approximate. It is a canonical, quasi-astronomical calendar rather than a true astronomical ephemeris. The Maya were aware that these tables, as related to the almanac, would not forever remain in synchronization with the celestial reality of eclipse occurrences. A correction procedure was required.

As a generalization, it does not seem that the Mesoamerican way was to use simple intercalculations, the occasional addition or subtraction of days to realign the tables (e.g., the leap year). They chose instead to use a unique base-shifting mechanism. Instructions were included with the tables to tell the reader of the codex how to change the base from which the calendar was calculated and how to truncate a cycle to realign it with astronomical reality while maintaining synchronization with the ritual 260-day Almanac. The solar year was always of secondary importance.

Though the correction mechanism for the Dresden eclipse tables is far from being understood, it seems to involve the three series of sacred almanac days displayed on the lower portions of most of the half pages. Each of the three rows of almanac days had a one-day regressive shift of base relative to the previous row. Because the "Palenque moon ratio" upon which the table is based is an approximation, the 405 average lunar months composing the table (a period of about 32 3/4 years) actually fall short of the total 11,960 days of the cycle by about one-ninth of a day. After nine repetitions of the table, almost 300 years, the accumulated error would cause the table to fall behind by about one day. In this same period of time, the node passage intervals (the lunar half-years) would fall short of the day count by about 14 1/2 days which is close to the ecliptic limits. This would suggest that more frequent corrections might have been necessary, perhaps every 33 years or so at the repetition of each eclipse cycle.

The arguments here are quite complex but have been well described by Lounsbury (1978). The most intriguing hypothesis proposed to account for the three almanac tables with their one-day relative base shifts rests on the fact that ten almanacs (2600 days) are just slightly over one day greater than an eclipse interval known to the Maya. If this 2600-day quantity is subtracted from any other eclipse cycle, there is just over one day less than another working eclipse cycle remaining. The application of these one-day base shifts at the correct times could neatly compensate for the accumulated error in lunar conjunction timings while *maintaining the same day* in the sacred almanac--a vital consideration for the Maya. The optimum application of this correction scheme gives a ratio of 4367 lunations to 128,960 days which yields an equivalent lunar month of 29.530570 days, substantially closer to the astronomically correct value of 29.530588 days than the "Palenque moon ratio" which is the basis of the table proper. The accuracy of this corrected ratio involves an error of only one day in about 4500 years!

The Maya Dresden Venus Calendar

Pages 46 to 50 of the Dresden Codex contain a ritual calendar of 37,960 days, or 104 years, which is based on the 584-day synodic period of the planet Venus. A sixth page contains the so-called correction table. Each of the five main pages represents one 584-day Venus cycle consisting of the planet's heliacal rise as a morning star, and disappearance at superior conjunction, reappearance as an evening star, disappearance at inferior conjunction. These four Venus stations were given the ritual lengths of 236, 90, 250 and 8 days, respectively, and are represented by the four columns at the left side of each page of the Codex.

At the time of its heliacal rise as Morning Star, Venus was seen as a malevolent deity whose rays could spear the unwary victim. The three pictures on each page represent, from top to bottom, the presiding Venus deity at heliacal rise, the spearing god, and the speared victim. The texts between the pictures speak of prophecies and dire consequences associated with these events. There are five pages in this calendar because five Venus cycles just happen to equal eight Vague Years,i.e., 5x584=8x365=2920 days. After eight years, Venus will appear at heliacal rise on approximately the same day in the yearly calendar. If we further require, as did the Maya, that those two periods also commensurate with the 260-day Almanac, we would produce a long cycle of 37,960 days (i.e., 65 x 584 = 104 x 365 = 146 x 260). This 104-year "Great Venus Cycle" is exactly two 52-year Calendar Rounds in length.

This need to commensurate Venus with the sun and the almanac explains the thirteen rows of almanac days that run across the top left sides of the pages. These give the almanac days for the four Venus stations in each cycle. We are instructed to go through the five pages thirteen times to obtain the full 104-year/65-Venus-cycle calendar.

The primacy of the 260-day calendar in the Venus table is exemplified by the fact that Venus always rises heliacally at the beginning of the table on a day named "1 Ahau" in the almanac. This day 1 Ahau was the mythical calendar name of Quetzalcóatl in his manifestation as Morning Star. To keep the beginning day of 1 Ahau and, at the same time, maintain alignment with the real Venus phenomena required a uniquely Maya correction technique.

The so-called correction table really contains two types of information. On the left side are numbers that relate the heliacal rise of Venus in historical times to an original "birth" of Venus near the end of the previous cosmological epoch. This is accomplished by first stating the 4 Ahau 8 Cumhu Calendar Round date for the beginning of the era. A "Ring Number" tells us to go back in time to a "Ring Number Base" in the last epoch. Then we are given a special distance number which, when added to this base, reaches the base date of the Dresden Venus calendar. It is highly significant that this distance number, which equals 1,366,560 days, is an elaborately contrived composite cycle filled with numerological significance. The date of the ring number base and the historical base are in the same position relative to the 260-day Almanac, the 365-day year, the 584-day Venus cycle, a 117-day cycle of the rain deity, the cycle of the 9 Lords and more. It is also 72 Calendar Rounds and thus 36 Great Venus cycles of 37,960 days. This number must have been designed to place the historical Venus events in harmony with the original mythological birth of the deity. It is precisely this type of relationship that has been discovered in the monumental inscriptions, where numerology links the dynasty to events surrounding mythological ancestral figures.

The *second* working portion of the correction table is on the right half of the page. This contains the scheme for shifting between three possible 365-day year-bases for the calendar and instructions as to when and where to truncate the calendar to realign it with reality. The true synodic period of Venus is 583.92 days. After five Venus cycles (eight years) the astronomical Venus would preceed the Calendar by 5 x 0.08 = 0.4 days. After the full 65 Venus cycles (104 years), the error would have increased to 5.2 days. John Teeple (1930) originally made a likely suggestion as to how the Maya used the correction tables. It involves a scheme of five separate truncations and base shifts over 301 Venus years in the amount of 24 days. If this be correct, it equates 301 Venus years to (310 x 584) - 24 days. This result is accurate to 0.08 day in 481 years, or about two hours in five centuries! The long-term accuracy of the Dresden Venus table is as striking as its short-term inaccuracy, and it shows that the tables were really to be used as a ritual almanac and not as an astronomical ephemeris for the planet Venus. Of necessity, the calendar-priests were obliged to provide long-term correction capability without sacrificing the all-important alignment with the sacred almanac. They seem to have accomplished their task remarkably well, and we must wonder if they realized how accurate their solution really was. Suffice it to say that the construction of this and other tables in the Dresden Codex contains artifacts from previously employed systems, showing that the Maya astronomers of the Classic period were tinkering with their calendars to produce better predictive capability and long-term accuracy. The eclipse tables and Venus calendar give us our only real glimpse at the intricacies of this process.

Conclusions

Mesoamerican astronomy was numerical or calendrical astronomy, not the more geometrical, positional astronomy that forms the basis of the Western tradition. Knowledge of true lengths of the celestial cycles must have been derived from pa-

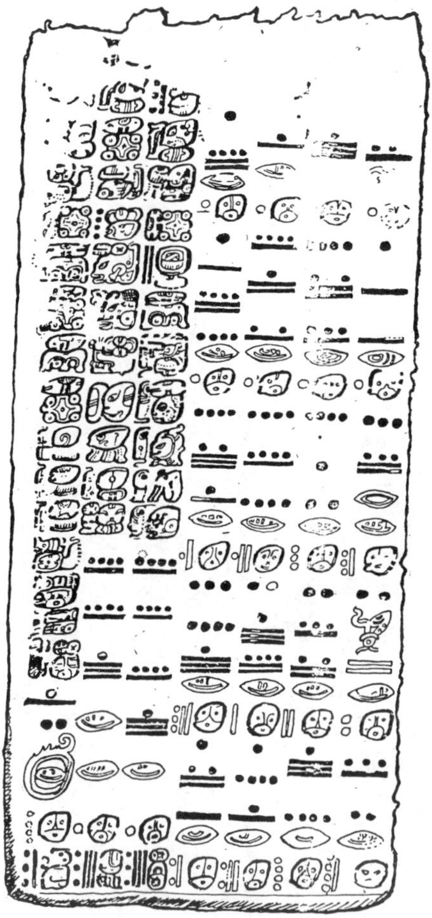

Dresden Codex page 24

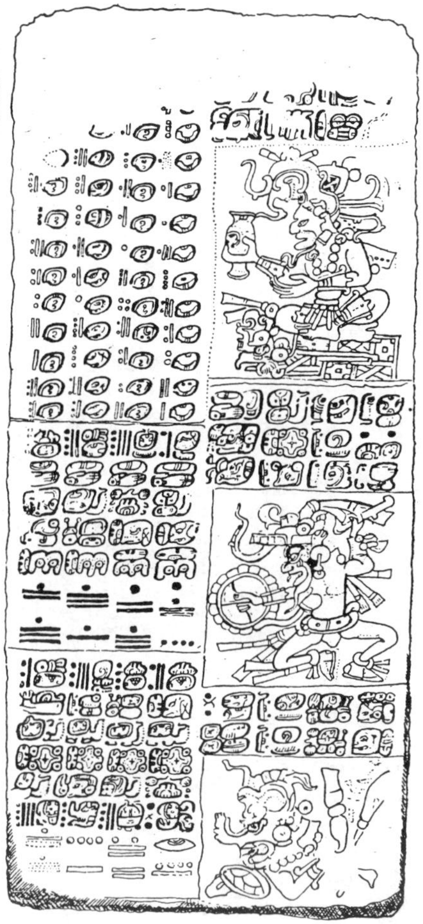

Dresden Codex page 46

The Dresden Codex also contains tables based on the observed cycle of the planet Venus, the sky's third brightest object. This table is actually a ritual calendar with a total of 37,960 days. This equals 65 584-day Venus cycles, but 146 260-day cycles and 104 365-day cycles (=1 "year") are both equal to the same total. These Venus tables begin on page 46 of the codex and continue through page 50. Page 24 contains references to a mythologically and cosmologically significant date in the distant past that is related to the date of the Venus calendar. The other half of this page incorporates numerical relationships that increase the congruence between the Venus count and the other day counts over long intervals of time. (after Villacorta and Villacorta, 1976)

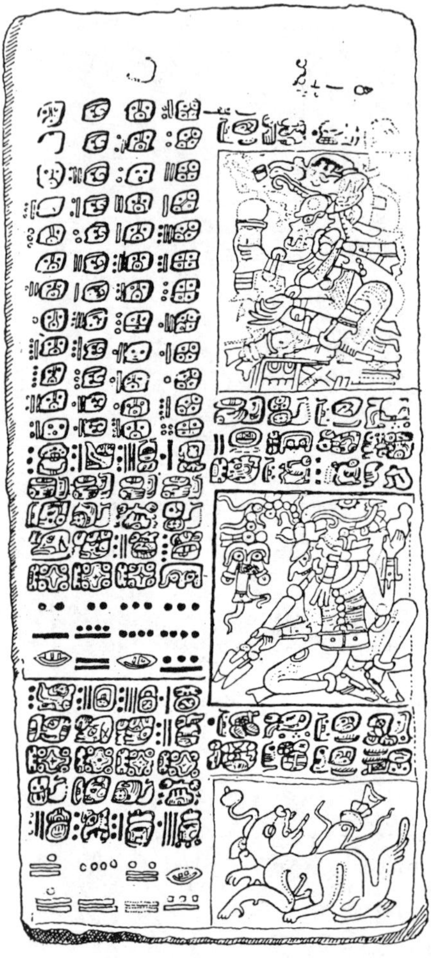

Dresden Codex page 47

Dresden Codex page 48

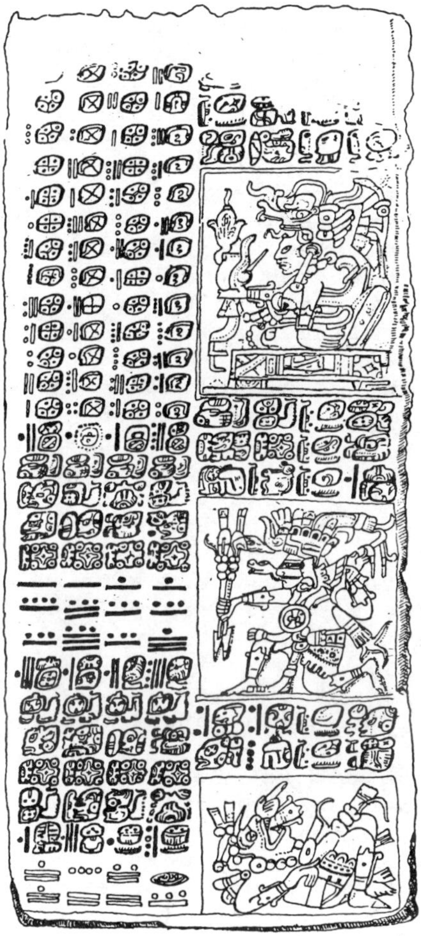

Dresden Codex page 49

Dresden Codex page 50

tient observation. Though zenith observations must also have been important, the bulk of astronomical practice was probably horizon-based and focused on heliacal and achronycal rise and set phenomena.

From all of the sources that we have at our disposal, it is clear that Mesoamerican astronomy was used to create and regulate ritual calendars and a deep-rooted system of cyclical prophecy. In an interactive system of mutual influence between Man and the cosmos, the role of the astronomer must have been to interpret both the regularities and irregularities of natural phenomena. We may suggest that the astronomers and calendar-priests were in a position of control of both the common man, through the almanac and agricultural calendar, as well as the elite. For the ruling dynasts, the astronomers controlled the selection of auspicious days for important royal events, and calculated the horoscopes and numerological ancestor-relationships. They would also have interpreted the omens and predicted the possibility of eclipses. We may also suggest (Carlson 1980) that the astronomer, calendar-priest, scribe, and historiographer functions were all closely related, perhaps even invested in the same person. If so, this would have given these individuals added power to manipulate "history" both before and after the fact.

Mesoamerican astronomy functioned not so much as science as we think of it but as a potent part of the cultural mechanism of social organization and control. Even though the record of this great civilization is largely erased, we can still get a glimpse of the individuals who observed the wonders of the heavens and tinkered with the elaborate mathematics and constructs of their world-order. Though we have travelled to the moon by virtue of the science and astronomy of our Western culture, we may still look at the Mesoamerican astronomical achievement with great respect and admiration as one of the most outstanding intellectual attainments of all time.

References

Aveni, Anthony F. (ed.)
1975 Archaeoastronomy in Pre-Columbian America. Austin: University of Texas Press.

Aveni, Anthony F. (ed.)
1977 Native American Astronomy. Austin: University of Texas Press.

Aveni, Anthony F.
1980 Skywatchers of Ancient Mexico. Austin: University of Texas Press.

Carlson, John B.
1976 "Astronomical Investigations and Site Orientation Influences at Palenque," (Merle Greene Robertson, ed.). Pebble Beach, California: Pre-Columbian Art Research, The Robert Louis Stevenson School, pp. 107-122.

Carlson, John B.
1979 "Numerology and the Astronomy of the Maya," in Archaeoastronomy in the Universe (ed. Ray Williamson). Los Altos, California: Ballena Press (Center for Archaeoastronomy, pp. 205-213.

Carlson, John B.
1980 "On Classic Maya Monumental Recorded History," (ed. Merle Greene Robertson). Third Palenque Round Table, 1978, Part 2, vol. V Austin: University of Texas Press.

Codex Dresden
1975 Codex Dresdensis: Sachsische Landesbibliothek Dresden Mscr. Dresd. R 310. Kommentar von Helmut Deckert und Ferdinand Anders. Graz, Austria: Akademische Druck- u. Verlagsanstalt,

Codex Fejervary-Mayer
1971 Codex Fejervary-Mayer: 12014, M, City of Liverpool Museums. Introduction by C.A. Burland. Graz, Austria: Akademische Druck-u. Verlagsanstalt.

Codex Grolier, and Coe, Michael D.
1973 The Maya Scribe and His World. New York: The Grolier Club, pp. 150-154.

Codex Madrid
1967 Codex Tro-Cortesianus (Codex Madrid). Museo de America, Madrid. Introduction and Summary by F. Anders. Graz, Austria: Akademische Druck-u. Verlagsanstalt.

Coe, Michael D.
1973 The Maya Scribe and His World. New York: The Grolier Club.

Collea, Beth A. and Anthony F. Aveni
1978 A Selected Bibliography on Native American Astronomy. Hamilton, NY: Colgate University.

Glass, John B.
1975 "A Survey of Native American Picturial Manuscripts", in Handbook of Middle American Indians (gen. ed. Robert Wauchope), vol. 14. Austin: University of Texas Press, pp. 3-80.

Goetz, Delia, Sylvanus G. Morley and Adrian Recinos (trans.)
1950 Popul Vuh: the Sacred Book of the Ancient Quiche Maya. Norman: University of Oklahoma Press.

Hunt, Eva
1977 The Transformation of the Humingbird: Cultural Roots of a Zinacantan Mythical Poem. Ithaca, NY: Cornell University Press.

Kelley, David Humiston
1976 Deciphering the Maya Script. Austin: University of Texas Press.

Leon-Portilla, Miguel
1963 Aztec Thought and Culture: a Study of the Ancient Nahuatl Mind. Norman: University of Oklahoma Press.

Leon-Portilla, Miguel
1973 Time and Reality in the Thought of the Maya. Boston: Beacon Press.

Lounsbury, Floyd G.
1976 "A Rationale for the Initial Date of the Temple of the Cross at Palenque," The

Art, Iconography and Dynastic History of Palenque, Part III (ed. Merle Greene Robertson). Pebble Beach, California: Pre-Columbian Art Research, The Robert Louis Stevenson School, pp. 211-224.

Lounsbury, Floyd G.
1978 "Maya Numeration, Computation and Calendrical Astronomy," Dictionary of Scientific Biography, vol. XV, suppl. 1, New York: Charles Scribner's Sons, pp. 759-818.

Maudslay, A.P.
1889-1902
Biologia Centrali-Americana; Archaeology. Vols. I-IV, London.

Morley, Sylvanus Griswold
1961 The Ancient Maya. (Revision G. Brainerd). Stanford, California: Stanfiord University Press.

Nicholson, Henry B.
1971 "Religion in Pre-Hispanic Central Mexico," Handbook of Middle American Indians (gen. ed. Robert Wauchope), vol. 10. Austin: University of Texas Press, pp. 395-446.

Proskouriakoff, Tatiana
1960 "Historical Implications of a Pattern of Dates at Piedras Negras, Guatemala," American Antiquity, vol. 25, no. 4, pp. 454-475.

Roys, Ralph L.
1967 The Book of Chilam Balam of Chumayel. Norman: University of Oklahoma Press. (First edition, Carnegie Institution of Washington, 1933.)

Ruz, Alberto
1970 The Civilization of the Maya. Mexico: Instituto Nacional de Antropologia e Historia.

Schele, Linda
1977 "Palenque: The House of the Dying Sun," (ed. Anthony F. Aveni). Austin: University of Texas Press, pp. 42-56.

Teeple, John E.
1930 "Maya Astronomy." Carnegie Institution of Washington Contributions to American Archaeology, vol. I, no. 2, publ 403, pp. 29-115, Washington, D.C.

Thompson, J. Eric S.
1971 Maya Hieroglyphic Writing: an Introduction (Third Edition). Norman: University of Oklahoma Press. (First edition, Carnegie Institution of Washington, 1950.)

Thompson, J. Eric S.
1972 A Commentary on the Dresden Codex. Philadelphia: American Philosophical Society.

Thompson, J. Eric S.
1974 "Maya Astronomy," Philosophical Transactions of the Royal Society of London, vol. 276, no. 1257 ("The Place of Astronomy in the Ancient World," ed. F.R. Hodson), pp. 83-98.

Tozzer, Alfred M.
1941 Landa's Relacion de las Cosas de Yucatán: a Translation. Papers of the Peabody Museum of American Archaeology and Ethnology, vol. XVIII, Cambridge, MA: Harvard University.

Villacorta C., J. Antonio and Carlos A. Villacorta
1976 Codices Mayas: Reproducidos y Dessarrollados (Second edition). Guatemala, C.A.: La Tipografia Nacional.

Anthony F. Aveni

6. The View from the Tropics

> "There is yet something else. Those people live like wild beasts on either side of the equator and between the poles men live like wild beasts, as Ptolemy says in his *Quadripartite*. And now this has been discovered by experience..."

When Scottish geographer John Major made that statement shortly after the European discovery of America, he expressed a total lack of understanding of another race's view of existence. Still, too often in our own age we tend to judge and evaluate the achievements of civilizations using our own present-day western standards as a yardstick.

The study of the heavens, as we shall see, provides ample evidence to demonstrate that if we are to understand an ancient people's view of the universe we must become acquainted with all aspects of their culture,and we must pay particularly careful attention to the environmental background in which they lived out their lives. How does a given system of astronomy work? How are celestial observations dictated by the needs of society? Why did such a system develop when and where it did? What can we learn about cultural behavior from it? These are the important questions. Is it science? Was it as good as western science? These questions are less important, even irrelevant, for they apply that familiar yardstick.

For tropically based civilizations a principle of geographic determinism operates. It gives people who live near "either side of the equator" a celestial outlook and orientation that is fundamentally

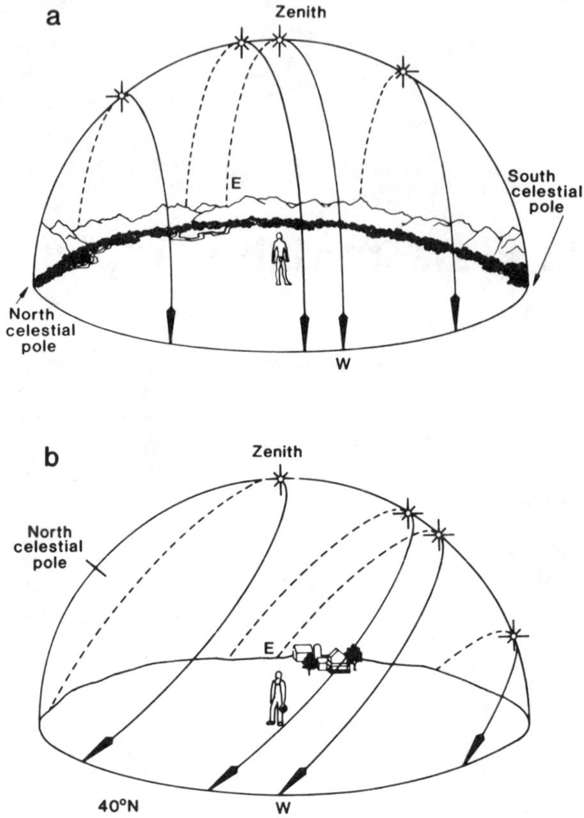

For an observer on the equator, the daily (or nightly) paths of celestial objects rise vertically from the eastern horizon. The celestial poles are on the horizon, and, accordingly, their importance is less here than at more northern or southern latitudes. Instead, the zenith, situated directly overhead, receives more attention, especially because the roving sun crosses through it on certain dates. At the equator, these solar zenith passages take place on the vernal and autumnal equinoxes. Elsewhere within the tropical zone the event still occurs, but on different dates. Beyond latitudes 23½ degrees north and south, no solar zenith passage is possible. An observer far from the equator sees a daily movement in the sky quite different than the view from the equator. The celestial paths are oblique to the horizon, and the celestial pole, now higher in the sky, assumes a more important identity. (Aveni)

different from, but often described as less sophisticated than, that developed by civilizations of higher latitudes. Tropical world views, once understood in their cultural and environmental context, are as rich and satisfying as our own.

Arorae is a small Pacific atoll in the Gilbert Islands. Its northern shore is dotted with half a dozen pairs of parallel rough-cut slabs, each about the size of a man, arranged horizontally and cemented into the ground. One pair is aligned to face the neighboring island of Tamana, 80 km distant; another points to Beru Island at 140 km. A third indicates distant Banaba, 700 km over the horizon. Islanders call them "Stone canoes" or "The Stones for Voyaging". They say the slabs were used to set directions for inter-island navigation. Each pair of stones aligns with the place where certain stars will appear or disappear at the sea horizon at different times of the night. For example, at sunset in August, the bright star Regulus aligns with the Tamana stone, while at midnight Arcturus gives the same bearing. The navigator simply memorizes a "constellation", which is really of a long chain of stars associated with the island he wishes to visit, and then steers his canoe toward them. In effect, the stars' rising positions are the points on a compass. The stone models on shore are used for instruction.

Such correlations for accurate long-distance sailing must have required generations of experience to develop. Indeed, as navigator David Lewis tells us in a recent book, the system still functions today. To solve the problem of landfinding, these maritime astronomers of Oceania developed two concepts that possess no analog in the cosmology and astronomy of the civilizations of the classical world: *linear constellations* and a *sidereal compass*. Both these indigenous creations, held in mind by oral tradition, seem to have been dictated by the environment. The situation and orientation of the sky in the tropics makes accurate navigating possible by this method, a technique that circumvents the objective magnetic compass and its accompanying point-by-point course plotting scheme.

In ancient societies, the sky and its contents lay at the very foundation of human cognition. Early hunter-gatherers, and later, more sedentary societies,

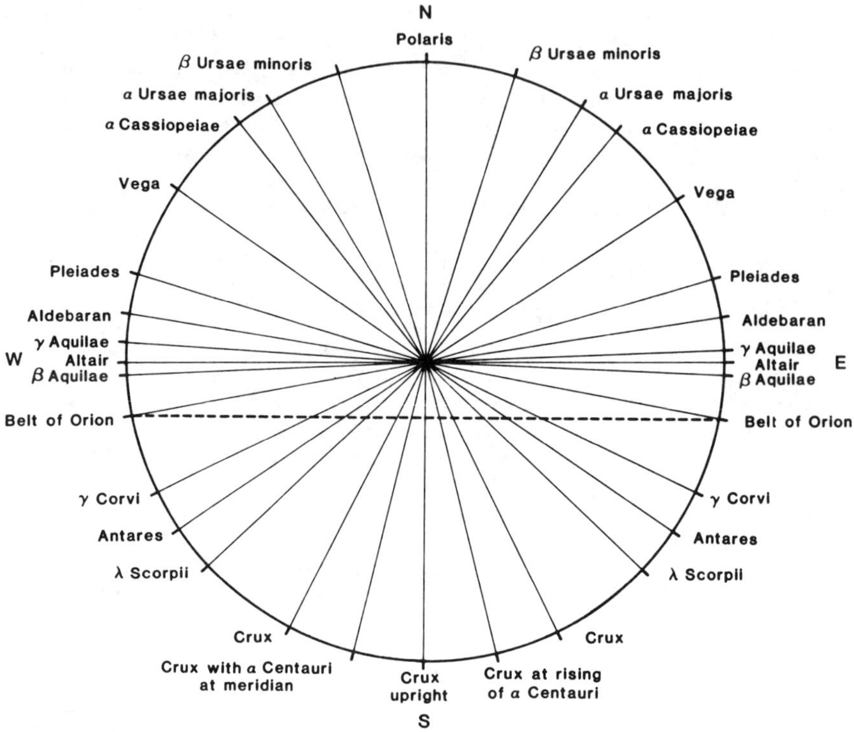

Caroline Islanders of the western Pacific, north of New Guinea, used the rising and setting positions of conspicuous and well-placed stars to establish directions and the bearings of various islands. (after Goodenough)

were profoundly influenced by the dependable precision of cyclic recurrence unfolding in the celestial canopy above them. To appreciate the difference between tropical and temperate celestial environments, let us imagine the earth's geographic coordinates to be reflected upon a hemispherical bowl that represents the sky. An observer at the equator sees the poles of the earth's axis of rotation on the horizon. The lines of latitude, projected onto the sky, pass vertically from eastern to western horizon. The rotation of the earth is mirrored in the movement of the stars and the motion is in the direction of these latitude lines. As a guide star rises over its associated island, the star may be aligned with the bow or mast of the ship and so indicate a fixed direction. When the star rises too high in the sky, another star conveniently appears to take its place. Away from the tropics the situation is quite different. There, star motion is a combination of both vertical and horizontal displacements. All celestial objects rise and set along oblique tracks, and the angle the star's path makes with the horizon grows smaller as the observer travels farther from the equator. Now consider the problem posed for a north Pacific or south Atlantic sailor who would adopt the system used at Arorae. As soon as his guide star appears, it begins to move laterally in relation to its place of first appearance. Unless a substitute star be immediately available, the navigator will deviate radically from his straight course. For example, if he is sailing from New England to Great Britain, within an hour he will be thrown off 10°, or 20 km along a 100 km segment of his course.

Practical-minded Oceanic peoples took a fact of geography and turned it to their advantage. They employed horizon-zenith astronomy because that system possessed obvious advantages when dealing with celestial motion. We intend to show in this article, through a wide selection of examples, that nearly all tropical cultures that developed indigenous astronomical systems, regardless of whether the function was largely practical or religious, gravitated toward a horizon-zenith system, an arrangement that stands in remarkable contrast to the ecliptic-pole or equator-pole systems developed by civilized nations of the ancient Classical world and the Orient.

According to Maud Makemson, an astronomer of a

generation ago who looked at Hawaiian astronomical concepts (latitude 20° N), sky watchers made practically no reference to the ecliptic and equator, but like most tropical people, they gave great importance to the cardinal points of the horizon. They imagined the sky to be supported by four pillars, each located at a cardinal direction. As the Maya, who called them the 4 Bacabs, they named these pillars in accordance with the viewpoint of an observer sitting at the east point of the horizon and looking along a fundamental east-to-west axis. North *(kukulu akau)* was the right-hand pillar (also meaning the direction "up"), while south *(kukulu hema)* was the left-hand pillar, or "down". In the neighboring Gilbert Islands (latitude = 30°S) the sky was segmented into named zones formed by slicing the celestial sphere several times in the vertical direction (parallel to the east-west line) and by cutting it with another set of lines running parallel to the horizon. The vertical lines were ridge poles (the great circle of the meridian), and rafters (small circles) and the horizontal arcs of the celestial sphere signified crossbeams or purlins. In this way, the location of a star could be described by placing it into one of a collection of imaginary altitude-azimuth boxes framing the celestial sphere. The orientation scheme of Egyptian astronomers (latitude 30°N) is similar to that of the Babylonians (latitude 32°N), while that of the Mursi of southwestern Ethiopia (latitude 5°N) is more like that of the Polynesians.

Inhabitants of ancient Mesoamerica (latitude 14° - 21°N) and their descendants also used astronomical alignments to monitor horizon events:

> Stone markers extending from behind Campo Santo up to top of high hill west of town. From Campo Santo to top approximately 1½ km. Sun rises on lines PS and DS observed from stones O and P on March 19, 1940, two days before the equinox. Sun rises this day at 6 degs 31½ ms. Direction observed with simple adjustable compass. Observations of the sun are made at the stone today by Zahorins for planting and harvesting.

I have used R.C.E. Long's transcription of field notes taken by anthropologist J. Steward

Lincoln concerning the astronomical practice of the Ixil Maya of the highland Guatemalan village of Nebaj (latitude = 15°N). From them we can reconstruct the scheme used by their ancestors, the great Maya astronomers of the Yucatán peninsula. It is a pity that Lincoln's premature death prevented him from studying this fascinating system in detail; however, his notes leave little doubt that these people carried on a tradition of marking precisely the location of sun at horizon from two different vantage points in order to time with precision the important planting dates in the tropical year. There are days, we are told, that are the only ones suitable for tilling the soil and sometimes the people work very fast in order to complete the planting on that day so that it shall not overlap with an unfavorable day.

The ancient Mesoamerican counterparts of these contemporary "horizon observatories" still remain in the monumental architecture and rock petroglyphs of Mexico and Central America, although the conscious motives that underlie many of the structural alignments still elude us.

At the wasteland along the Tropic of Cancer near Zacatecas, Mexico, we find an ingenious scheme still intact in the archaeological record. It consists of a double solar alignment, the sunwatchers stationed 7 km apart in a generally north-south direction. One observer, standing in the "Temple of the Sun" at the ruins of Alta Vista (built about 400 A.D.), can see the sun rise at the spring and autumn equinoxes over Picacho Montoso, a pinnacle located due east of the ruins at range 15 km. On the summer solstice an observer standing on Cerro El Chapin to the south watches the sun rise over the same peak. Moreover, an unusual rock carving marks the spot on Cerro El Chapin. A double circle quartered by a pair of intersecting lines is pecked into a flat rock on the edge of the summit of the hill. The design is identical to a family of petroglyphs found carved on rocks and in the floors of buildings at the ceremonial center of Teotihuacán, 650 km to the south. Furthermore, a similar marker appears in the floor of Structure A-V at the Maya ruins of Uaxactún, situated 2000 km away from the Tropic site.

These pecked cross-circles, which the author

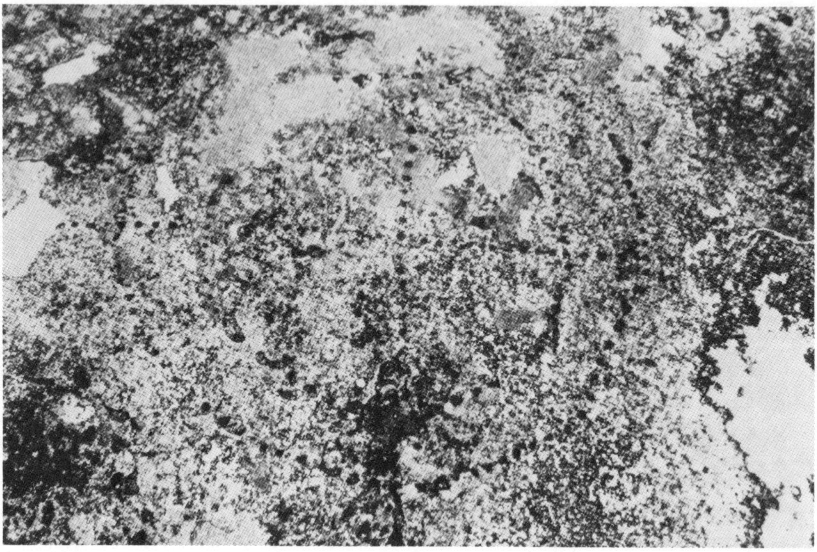

a

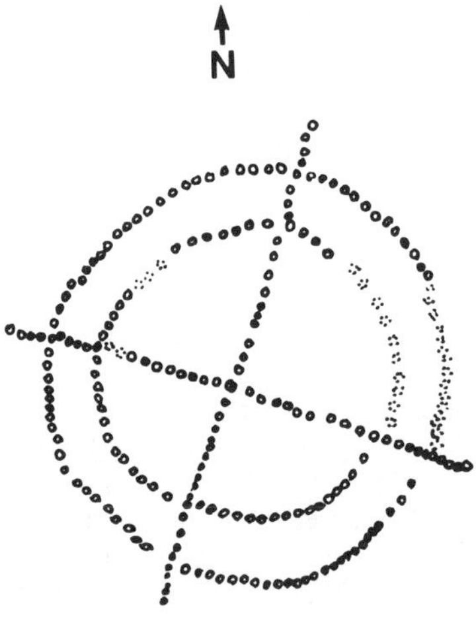

b

A room in the structure known as the Viking Group at Teotihuacán has this quartered circle, or pecked cross, design punched into its stucco floor. These symbols appear to have astronomical and calendrical significance and are found from the far north of Mexico to the Guatemalan jungle. ([*a*] Robin Rector Krupp; [*b*] Griffith Observatory, after Aveni)

a

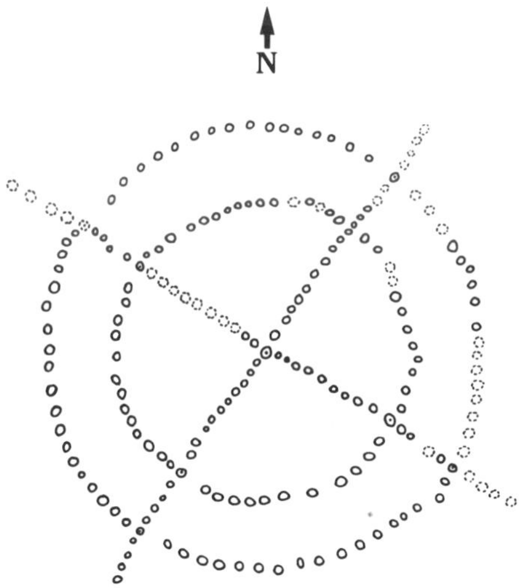

b

There are several other pecked crosses in the vicinity of Teotihuacán. This one is on the slope of Cerro Colorado, 3 kilometers west of the Viking Group, whose pecked cross it closely resembles. ([*a*] Robin Rector Krupp; [*b*] Griffith Observatory, after Aveni)

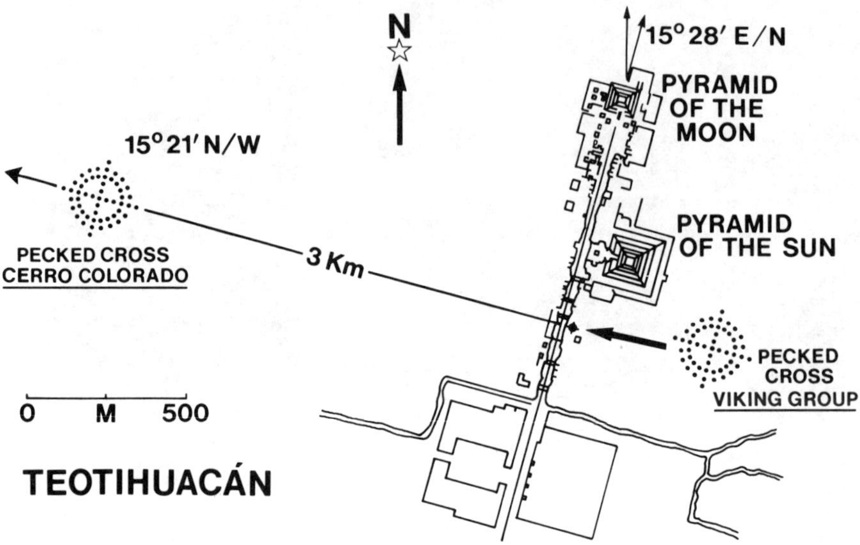

The pecked crosses of the Viking Group and Cerro Colorado establish a line almost exactly perpendicular to the main axis of Teotihuacán. Whatever else these symbols may mean, here they seem intended to preserve the fundamental orientation of the city. The line to Cerro Colorado also coincides with the setting of the Pleiades during the time of Teotihuacán's early development, and the Pleiades, in turn, may have acted as the herald of the year's first solar zenith passage here in central Mexico, at 19°42' north latitude. (Griffith Observatory)

and Horst Hartung, an architect at the University of Guadalajara, have mapped and measured in detail, are found to exhibit calendric as well as orientational properties. In many cases, the Mesoamerican 260-day calendar (a cycle consisting of the alternation of a series of numbers, from one to thirteen with a sequence of 20 named days) is incorporated into the design in terms of the tally of the peck marks. Indeed, the pecked circles resemble the "calendar wheels" found in the codices and post-Conquest chronicles. The cross-circles seem to have been devised by astronomers of the great Teotihuacán empire, an advanced civilization that made its presence felt throughout Mesoamerica during the millennium of its existence around the beginning of our Christian era. Fifteen pecked cross markers are found in the valley of Teotihuacán, northeast of modern Mexico City. A pair of them was probably used to orient the rectangular grid plan of ancient Teotihuacán by alignment with the setting point of the Pleiades star cluster.

Archaeologist J. Charles Kelley, in charge of the excavations of Alta Vista, has produced evidence, both in the form of pottery and radio carbon dates, to suggest that Alta Vista is a site with considerable Teotihuacán influence (late Xolalpan phase, circa A.D. 400). The situation and orientation of the site and the site's relationship to its environment suggest horizon astronomy must have been a factor in the grand plan. The astronomers seem to have been seeking the place where the sun turns back from its migration north (what we would call the latitude of the Tropic of Cancer). At that place the sun will stand perched in the zenith at noon on the longest day, the first day of summer. A shadowless moment occurs as the sun arcs over the zenith and returns to its southern realm. Our alignment studies suggest that the people of Alta Vista used their landscape to demarcate the year by the equinoxes and the summer solstice. Though we have yet to find one, an observing station for the winter solstice may well lie to the north of Alta Vista. From there the astronomer could watch the sunrise over Picacho Peak at the winter solstice.

Evidence from another quarter supports the horizon scheme as we have outlined it. The studies of field anthropologist Ruth Bunzel allude directly to such observing techniques among the Zuñi. She

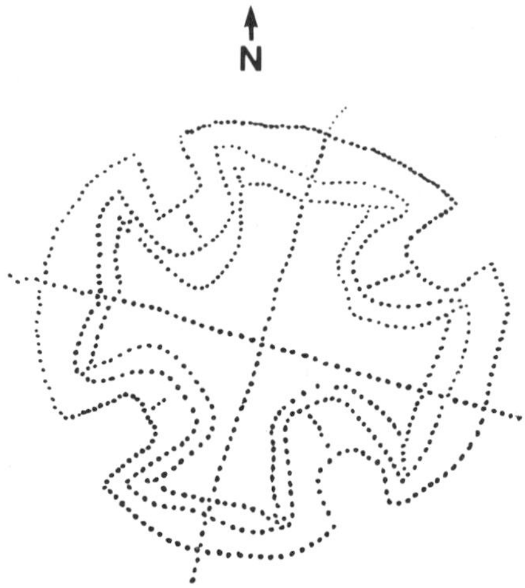

a

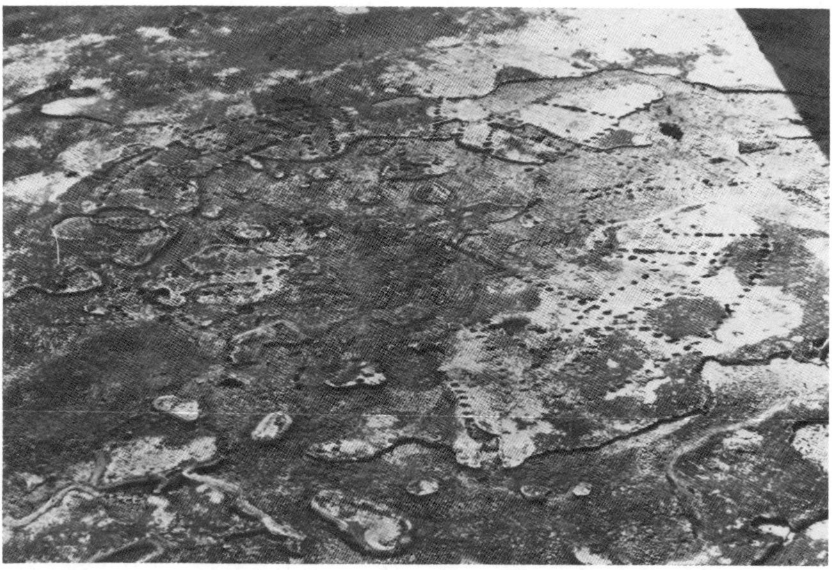

b

c

The number 260 is important as the number of days in one Mesoamerican calendar cycle, the so-called Sacred Almanac, and 260, or a number close to it, shows up in a number of the pecked crosses. One of them [*a*], at Teotihuacán, has 260 marks around its perimeter, and in shape it resembles a design in the central Mexican Codex Fejérváry-Mayer (illustrated elsewhere in this book) and a similar design in the Maya Codex Madrid [*c*]. The cross-shaped borders in both of these codex drawings comprise 260 dots. ([*a*] Griffith Observatory, after Aveni; [*b*] E. C. Krupp; [*c*] S. Peet, Myths and Symbols or Aboriginal Religions of America, 1905)

The Hall of Columns at Alta Vista has its corners oriented to the cardinal directions. Here we are looking from the south corner to the north. (Robin Rector Krupp)

A walkway (?) breaks off from the Hall of Columns toward cardinal east and toward Picacho Montoso, the distinctive peak that marks the position of equinox sunrise. (E.C. Krupp)

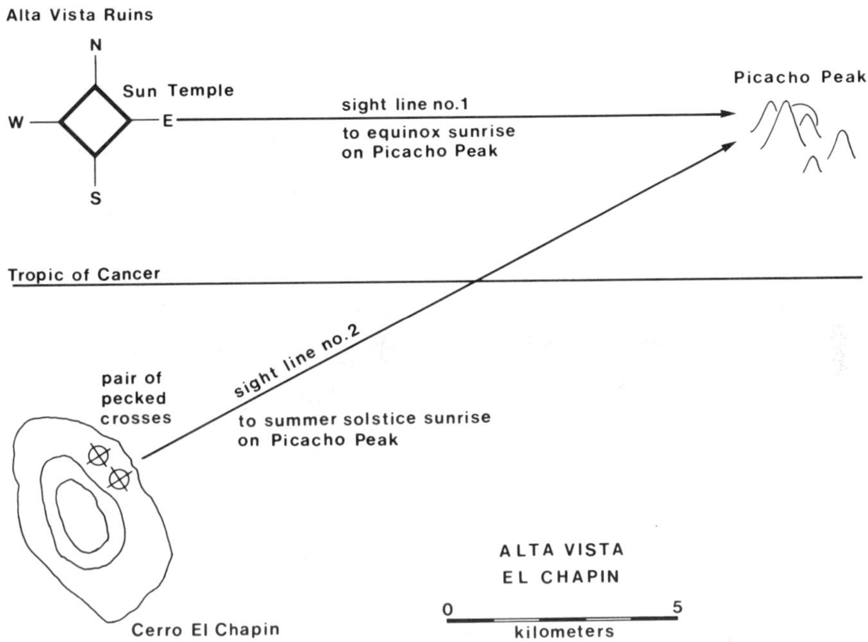

Near Alta Vista, on Cerro el Chapin, a pair of pecked crosses marks the backsight for observation of the summer solstice sunrise over Montoso Picacho, the same foresight for the equinox at Alta Vista not far to the north. Here, on the Tropic of Cancer, there is only one solar zenith passage, and it occurs on the summer solstice. Farther north, solar zenith passage is impossible. (Griffith Observatory)

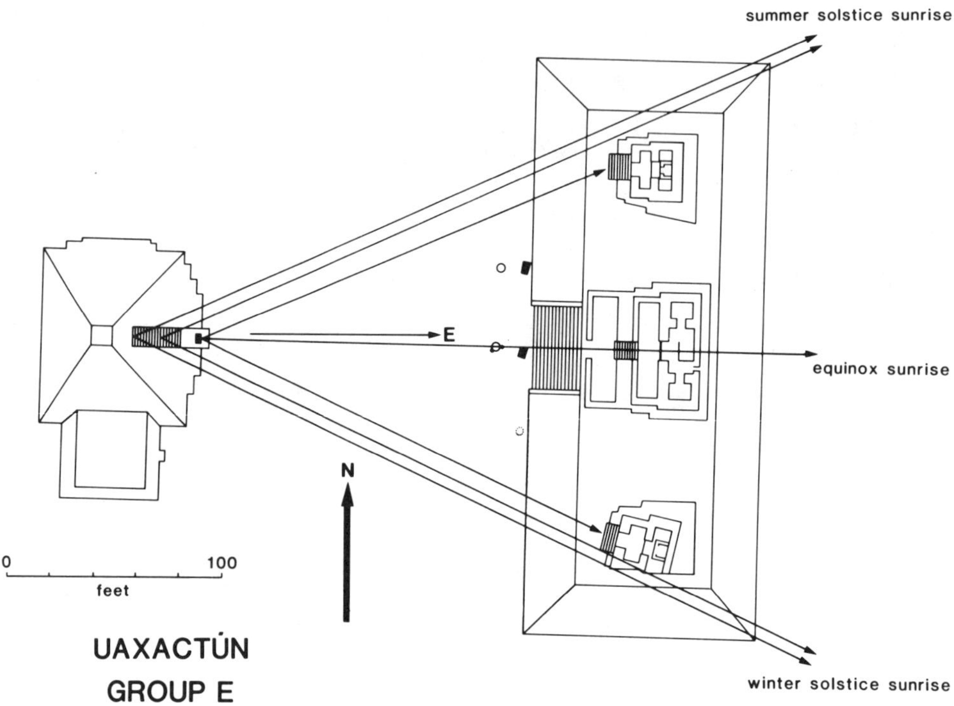

A similar interest in the solstice and equinox sunrises may be expressed in the architecture of this group at Uaxactún, a Classic Maya site in Guatemala (Griffith Observatory, after Hartung)

tells us that the winter solstice starts "when the sun rises at a particular point on the mesa to the southeast". A similar statement applies to summer solstice. This sort of documentation from cultural anthropology serves as a check on our astronomical reasoning derived from alignment data acquired at the archaeological ruins with the surveyor's transit.

The astronomically oriented buildings of Mexico are even more numerous than the pecked circles. The Caracol of Chichén Itzá (latitude 21°N) and the Palace of the Governor at Uxmal are both aligned with Venus horizon events. We might anticipate that the flat featureless horizon would necessitate the incorporation of any significant astronomical directions into works of standing architecture, and this is exactly what we find.

There is good evidence from the Spanish Chronicles that the Inca of Peru (latitude 13½°S) erected sets of vertical pillars to follow the sun's motion, at least over the western horizon of Cuzco. The socio-political organization of this city was arranged according to the *ceque* system, a network of 41 radial lines emanating from the Temple of the Sun (called the *Coricancha*) located in the middle of the Cuzco valley. The Spanish chroniclers Cristobal de Albornoz and Bernábe Cobo, writing only a few generations after Pizarro's penetration into Peru, provide a detailed account of the directions of the ceque lines. They describe and locate in the valley of Cuzco 328 *huacas* or sacred places that lie along the ceques. Different kinship groups controlled and cared for the huacas of their ceque, and they worshipped at certain huacas on particular days of the year for different purposes. The system constituted a socio-economic map of the city that also provided ritual information to the citizenry. From studies of place names in the Cuzco environment, anthropologist R. T. Zuidema of the University of Illinois and the author have been able to locate and map approximately one-quarter of the ceque lines. We find that astronomical information is encoded in portions of the system. Three ceque lines are of particular interest to those of us studying horizon-zenith astronomy. Cobo explicitly states that observation towers or pillars were located along these ceques. He identifies the ceques by number within a given quadrant of the city. For example,

The Caracol of Chichén Itzá, Yucatán, could have been used to observe the planet Venus and perhaps other celestial objects. (E. C. Krupp)

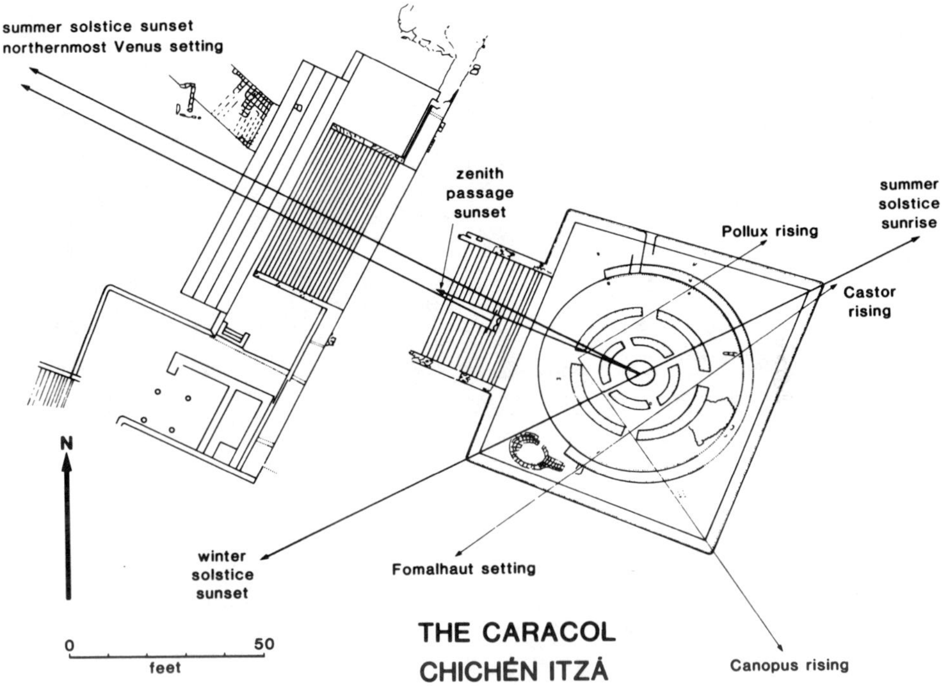

Numerous astronomical alignments conform to the orientations of the upper and lower platform of the Caracol. These alignments may have been symbolic references to the celestial objects that interested the Maya of Chichén Itzá. (Griffith Observatory, after Aveni, Gibbs, and Hartung)

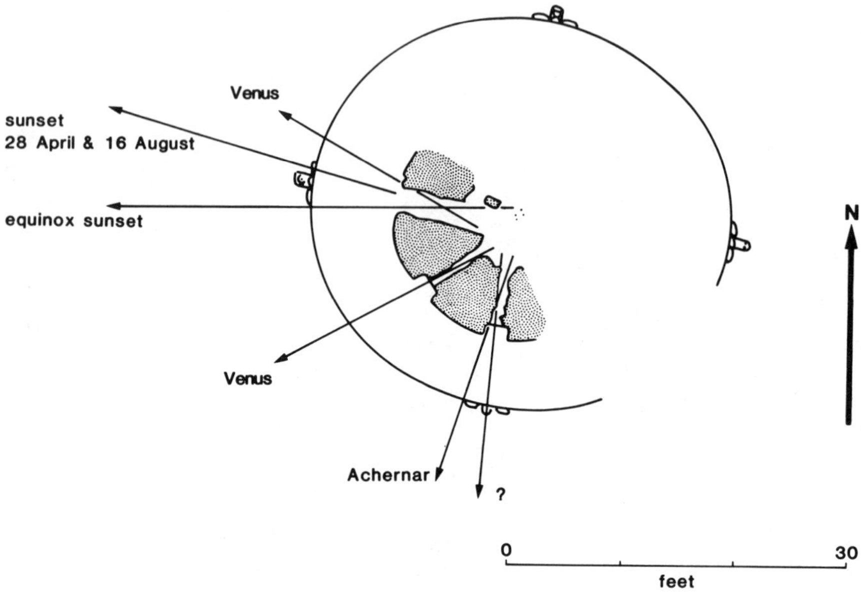

THE CARACOL, CHICHÉN ITZÁ

In the course of its excursion, Venus reached the extremes indicated by the cross-jamb alignments through windows of the Caracol's upper tower. (Griffith Observatory, after Aveni, Gibbs, and Hartung)

<u>Facing page</u>

One of the more impressive astronomical alignments at Uxmal, a Classic Maya ceremonial center in the Puuc region of Yucatán, coincides with the most southern rising of Venus. From the central doorway of the Palace of the Governor this line extends southeast across a cylindrical stela, a small platform, and on to the ruins of Nohpat act as a Venus foresight on the otherwise flat horizon. (Griffith Observatory, after Hartung)

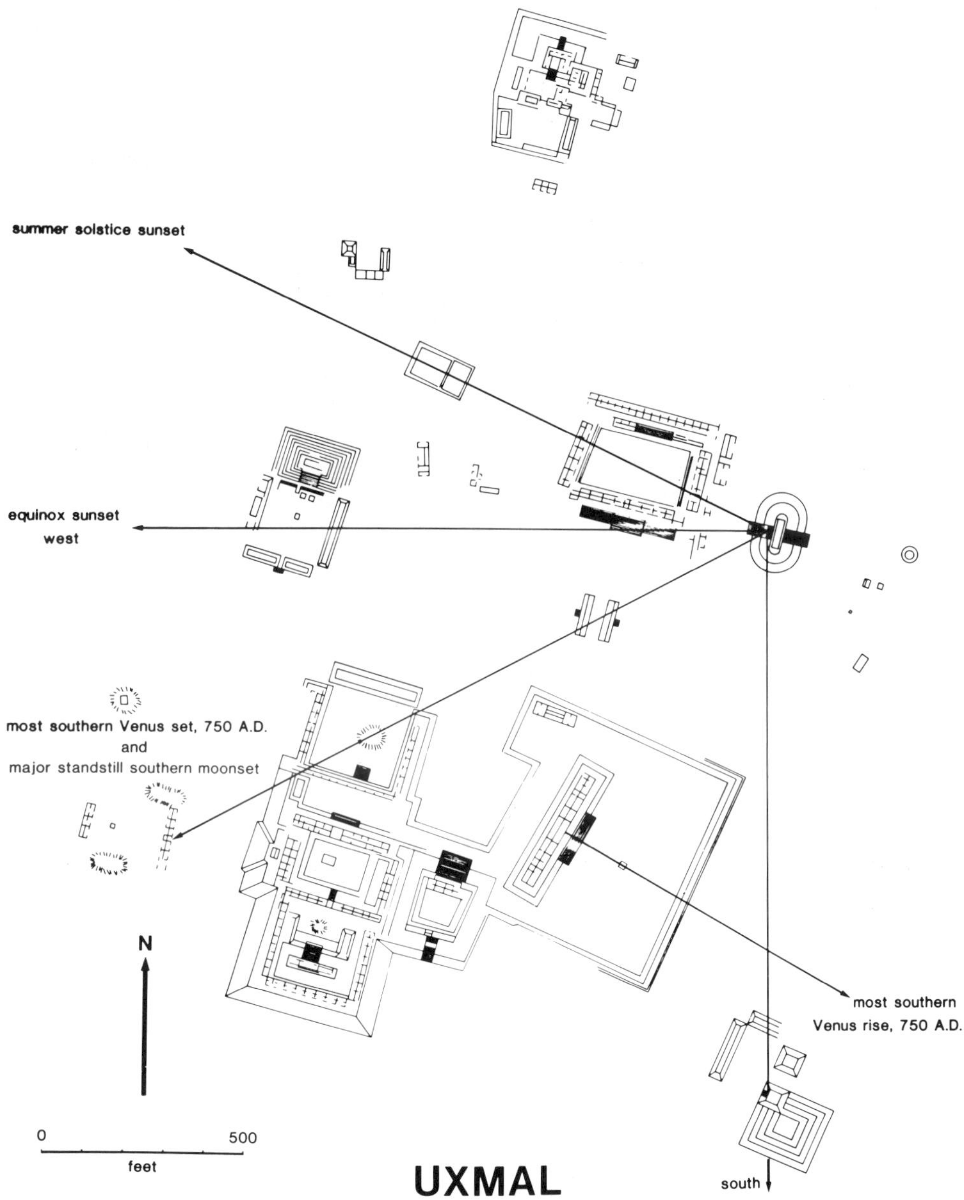
summer solstice sunset
equinox sunset
west
most southern Venus set, 750 A.D.
and
major standstill southern moonset
N
0
500
feet
most southern
Venus rise, 750 A.D.
south

UXMAL

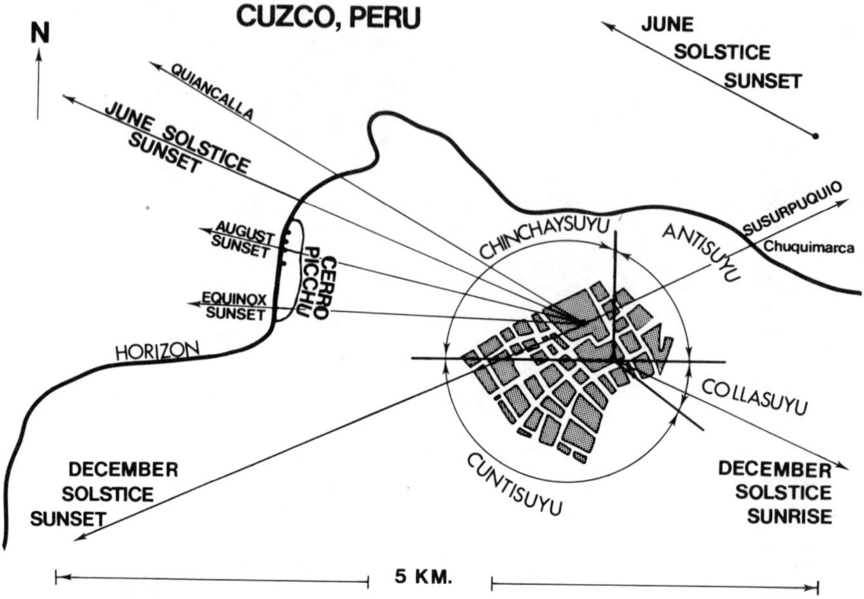

Cuzco, the Inca capital, was subdivided into four "quarters", as was the entire empire. Foresight towers for a solar observatory once stood on Cerro Picchu to the west and marked the positions of August sunsets, important in the agricultural calendar. (Griffith Observatory, after Aveni)

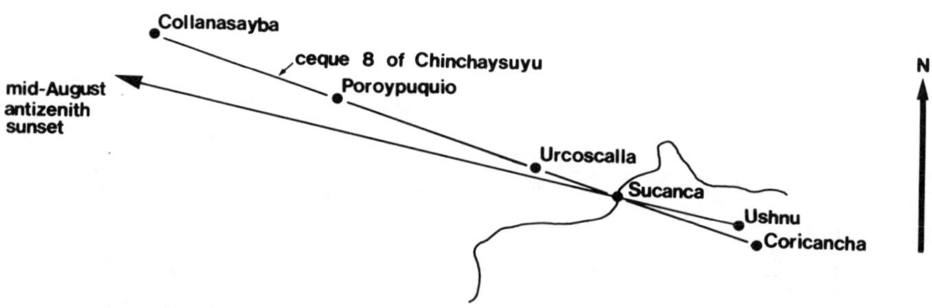

The location of the Cerro Picchu Towers (Sucanca on this plan) was reconstructed by tracing out ceque 8 of Chinchaysuyu quarter with descriptions of its shrines (*huacas*) in the chronicles of B. Cobo. Assuming the key date was the time of anti-zenith passage of the sun (a temporal inversion of the zenith event)--a possibility consistent with local tradition--the backsight was pinpointed. This was an important monument known as the Ushnu. (Griffith Observatory)

> Cuntisuyu Quadrant - ceque 8, huaca 7: "Sucanca was a hill where the irrigation canal of Chinchero comes. On it were two pillars or monuments to signal that when the sun arrived there it was time to begin to plant maize. The sacrifice made there was addressed to the sun and they asked that it would arrive on time so that they would have good reason to plant..."

Cobo's other statements imply that a pair of structures was designed to frame the sun's disk at each of the solstices, but other remarks indicate that four pillars were used to mark the sun during a critical series of sunset observations taken about the middle of August. Another purveyor of Inca history, the Anonymous Chronicler in Maurtua, tells us why this was so,

> High on a hill over-looking Cuzco from the west lay a grouping of pillars and when the sun passed the first pillar they prepared themselves for planting in the higher altitudes, as ripening takes longer there.
>
> When the sun entered the space between the two pillars in the middle it became the general time to plant in Cuzco; this was always in the month of August.
>
> And when the sun stood fitting in the middle between the two pillars they had another pillar in the middle of the plaza, a pillar of well worked stone about one estado high, called the Ushnu, from which they viewed it. This was the general time to plant in the valleys of Cuzco and its surroundings.

Evidently the northernmost pillar served as a warning device that the planting season in the Cuzco valley was approaching, as was sunset on its north-to-south trek along the horizon. People who cultivated crops at higher altitudes, where growth occurs at a slower pace, were allowed sufficient additional time to sow their seeds before planting commenced in the valley. Although the chroniclers provide no specific evidence, we speculate that the southernmost pillar may have been used as a similar warning device during the harvest season when the sun, moving from left to right, or south to north,

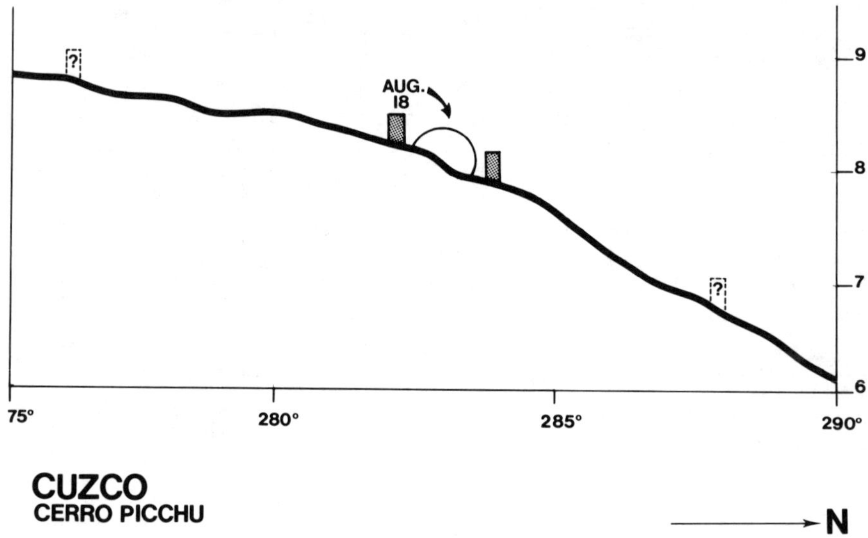

From the chroniclers' accounts, we surmise how the sun's position with respect to the towers of Cerro Picchu alerted the Inca to the appropriate planting times. The emphasis on the 18 August, solar anti-zenith passage, date is consistent with the emphasis placed by tropical peoples on the vertical, zenith-nadir axis. (Griffith Observatory, after Aveni)

passed it in April. Using the chroniclers' estimates of the distances separating the pillars we find that a month of sunsets would traverse the entire span of the pillars. With the sun centered in the space between the inner pillars on August 18, the day it passes directly under the zenith-nadir axis, we found that it would arrive at the northernmost pillar on August 4, one-half of a lunar month before mid-passage and at the southernmost pillar on August 31, one-half month after. This raises the interesting question of whether observations of the moon might have been correlated with the solar sightings. Bernábe Cobo says there were sucanca to signal the months of the year, apparently, when the sun arrived at successive pillars. Because the method is barely outlined by the chroniclers, we have difficulty interpreting their confused statements about the moon. Sadly, the old historians paid too little attention to the sophisticated calendar system that was written in the landscape of Cuzco, under their very feet.

Over three field seasons in Cuzco, Zuidema and I were able to pinpoint the original locations of the astronomical pillars by seeking out neighboring huacas and ceque lines according to their place names, as given by the chroniclers and by local informants. Nothing remains at any of the sites now, for the Spanish, ignorant of the Inca cosmic view, had dismantled the structures before 1600 and used the stones as building materials. Nevertheless, it would be worthwhile for archaeologists to excavate to see if any sub-structures of these important calendric indicators remain.

When we followed the astronomical directions back into Cuzco from each group of pillars where the sun was observed to set, we were surprised that only one of the three directions led to the Temple of the Sun, the other two ending at observing stations to the north.

Three different observing stations each used a set of pillars on the western horizon as a foresight--two pillars in each of two sets with a third set composed of four pillars. This is hardly a result we might have anticipated. But then, Inca astronomy is not like Classical astronomy. The political structure of the empire was visibly interlocked with astronomical thought. The city of Cuzco was

We know such towers existed, and we have some idea what they looked like. This illustration from Don Felipé Huamán Poma de Ayala's *Nueva Coroñica Buen Gobierno*, written between 1567 and 1615, depicts the mojonadors in the act of measuring stone blocks and setting them in place. A finished tower appears in the background. (Royal Library of Copenhagen)

divided into a northern and southern sector, called Hanan-Cuzco and Hurin-Cuzco, respectively. Everything that transpired in the northern half of the domain was governed or cared for by the people who inhabited that ward of the city. The same held for the south, and their duties evidently included even the procurement of astronomical observations along the jagged horizon that framed their isolated world.

If there are alignments to the sun at horizon in the tropics, might we also expect orientations to the sun at zenith, the fundamental reference pole? Here the evidence is quite clear not only in Cuzco but also in Mesoamerica and Malaysia.

While the principle of the gnomon, or vertical pole, to cast shadows is well known in the Greek and Roman sundials of the Classical world, for the people of ancient Java (latitude 7°S) the device takes on a wider significance. There it is used not only to divide the space of the sky but also to partition the time of the year.

The cruciform plate of the Javanese dial is fashioned to reckon the start of the year from the June solstice, the first day of winter in the southern hemisphere. The so-called rustic year is segmented into 12 months of unequal duration, ranging from 43 days (the first, sixth, seventh, and twelfth months) to 23 days for certain months in between. Alfred Maass, a Dutch ethnohistorian, demonstrated that these unequal time intervals are precisely obtainable if one uses as the guiding principle not equal units of time but equal units of *length* traveled by the shadow cast by the tip of the gnomon at noon. The passage of sun through zenith serves as the pivotal point for dividing up the year.

In the latitude of central Java a very special condition occurs in gnomonic geometry. When the sun stands on the meridian at noon north of the zenith on the June solstice, the shadow length, measured to the south of the base of the vertical pole is exactly one-half the length of that which obtains when the sun at noon lies at the December solstice south of the zenith, then projecting a shadow to the north. Javanese chronologists halved the shorter segment and quartered the longer, thereby producing a 12-month calendar. As it often occurs in most cultures,

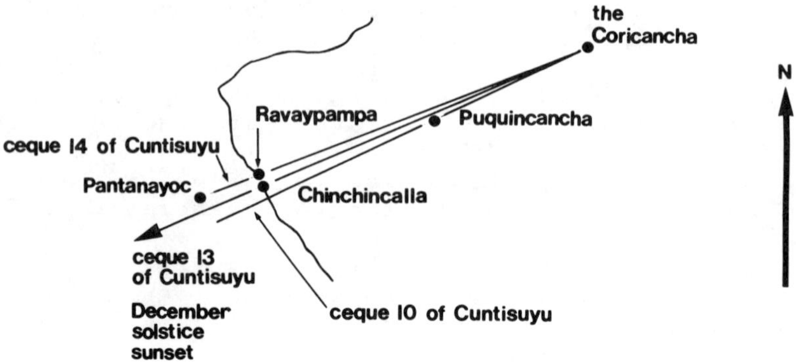

A complex system of imaginary lines delineated by shrines (*ceques* and *huacas*) embodied the organizational pattern of Cuzco. Some of these lines--but not all of them--are astronomically oriented. Ceque 13 in the Cuntisuyu quarter points to the December solstice sunset. (Griffith Observatory)

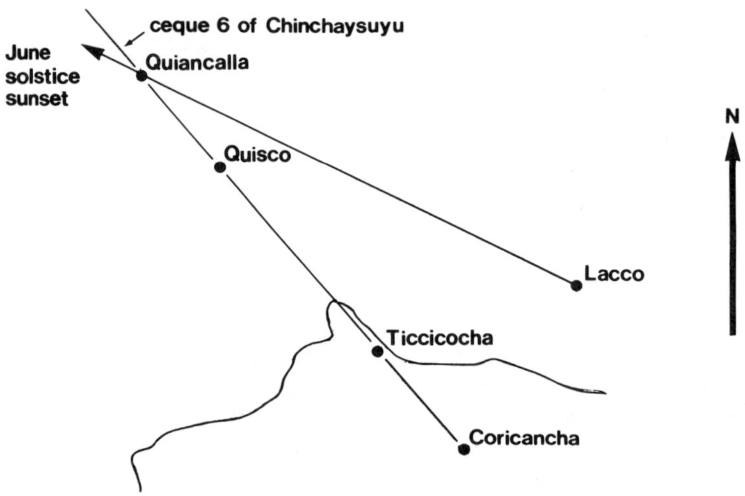

All of the ceques radiated from the center of Cuzco, the temple known as Coricancha, but not all of the astronomical sightlines described by the chroniclers originated there. By tracing ceque 6, Chinchaysuyu quarter, to a known solstice sunset foresight (Quiancalla) and working back along the solstice line southeast, the backsight for the Inca observatory was pinpointed at Chuquimarca, a sun temple near the carved rock Lacco. (Griffith Observatory, after Aveni)

the number 12 probably was chosen to integrate full moons into the yearly scheme. Because the sun moves more slowly at the solstices, we should expect the first and last months along with the sixth and seventh months, when the shadow approaches the southerly and northerly extensions, respectively, to be those of greatest duration. Months defined by equal shadow lengths, when the sun is near the equinox, ought to be rather brief. This is precisely what occurs. Thus, the Javanese time scheme yields months with peculiar lengths such as 25, 27, 41, and 43 days. Calendar keepers later altered these intervals by one or two days either way in order to make them coincide more accurately with agricultural practices.

The architects of this curious calendar took a spatial unit and transformed it into a temporal one. But the Malaysian attempt to fuse space and time is not unique to that culture. This notion also appears in the calendars of ancient Mexico, where the day names of the 260-day calendar are assigned positions around the horizon and the year-bearers or names of New Year's day in succeeding tropical year calendars are associated with different cardinal points. If we think such ideas foreign, we might wonder whether the attempt to wed the elusive concept of time to the more concrete realm of space is not still alive in our own culture in the world view of Einstein and the relativists of modern physics.

Garcilaso de la Vega, the romantic historian of the Inca, leaves no doubt that the ancient Cuzqueños of 15th century Peru used the gnomon:

> When the priests felt that the equinox was approaching, they took careful daily observations of the shadows cast by columns. The columns stood in the middle of a great circle filling the whole extent of the mesa or courtyard. Across the middle of a circle a line was drawn from east to west by a cord, the two ends being established by long experience. They could follow the approach of the equinox by the shadow the column cast on this line, and when the shadow fell exactly along the line from sunrise and at midday the sun bathed all sides of the column and cast no shadow at

> all, they knew that that day was the equinox. They then decked the columns with all the flowers and aromatic herbs they could find, and placed the throne of the Sun on it, saying that on that day the Sun was seated on the column in all his full light.

Though Garcilaso has confused the equinoxes with the zenith passages in his discussion, he gives us the impression that the Inca had developed a sophisticated scheme for timing the solar year via the horizon system. In addition to horizon pillars, shadow casting columns existed throughout the Inca empire, which stretched over 30 degrees of latitude in a narrow north-south strip.

A set of columns near Quito, Ecuador (latitude 0°17'S), was held in special veneration, says Garcilaso, for there the sun is "in plumb line" at the equinoxes. "...They afforded the sun the seat he liked best, since he sat straight up and elsewhere on one side." But archaeological evidence bearing on the pillars is difficult to establish, for, as Garcilaso also tells us, "...like the horizon pillars of ancient Cuzco they were very properly pulled down and broken to pieces by the Governor Sebastian de Balalcazar, because the Indians worshipped them idolatrously. The others throughout the empire were demolished by the Spanish captains as they came across them."

In Mexico, unlike in Java or Peru, where Garcilaso de la Vega alludes to it, there is little evidence for shadow casting, but we do know the sun was observed at zenith, and so were the Pleiades. Next to a diagram of Aztec constellations drawn by a local informant, Padre Bernardino de Sahagún, the historian of early Mexico City, tells us:

> These are representatives of the Pleiades which mark the 5th cardinal point. At the beginning of a period of 52 years, fire was newly kindled when the Pleiades were in the zenith at midnight. The flaming up of this fire was a sign to the anxious waiting multitude that the world was not, as they had feared, to be swallowed up in darkness but that a new era would be granted to mankind.

The 5th cardinal point was the zenith and the other four lay at right angles to it in the horizon plane. Perhaps we now understand why the Pleiades were so important to the Teotihuacanos.

In Yucatán, Juan Pio Perez, a 19th century chronicler of the town of Peto, writes that the year there began at zenith passage, which was the 16th of July, but he gives no details of how the date was determined. While there is no direct argument to suggest that gnomons were used, there is some rather good evidence to illustrate how they marked the shadowless moment. We have identified two peculiar architectural structures in Mexico as "zenith tubes", one at Monte Albán (in the Zapotec highlands near Oaxaca, latitude 17°N) and the other at Xochicalco (south of Mexico City at latitude 19°N). These tubes seem to have been intended to admit the sun's light directly and to pass it vertically into a darkened chamber.

When we supplement the discovery of the two examples of zenith tubes in Mexican architecture with Zuñi ethnologist, E. C. Parson's statements about the priests' observationsof the noon-time sun through a hole in the roof of his kiva, we can begin to draw some preliminary conclusions about how the zenith sun and the Mesoamerican calendar were connected. Zenith tubes, rather than the gnomons more familiar to our culture, constitute the most recognizable method for watching the sun traverse the 5th cardinal direction. Furthermore, we find no better hypothesis to account for the existence of the vertical tubes so carefully engineered into the architecture at Xochicalco and Monte Albán. We might well look for other examples of this unusual device in the archaeological remains of the civilizations of the tropical Americas.

To return to where we began, zenith *stars* instead of the zenith sun were important in Polynesia and again the motivation was navigation. Because the declination (the angular distance north or south of the celestial equator) of a star that passes the zenith is the same as the latitude of the observer, every island could be associated with its own zenith-passing guide stars. Though he probably conceptualized it in different terms, we could say that the navigator would know he reached the parallel of geographic latitude of his

The masonry opening of a narrow vertical shaft at Xochicalco is 17 feet above the ceiling of a cave chamber cut into the rock below. This tube and room may have been used to monitor zenith passages of the sun. (E. C. Krupp)

destination when the arc of a guide star's course crossed the overhead position. For this reason Sirius was the guide star for the Fiji Islands (latitude 17°S) and Rigel, in our constellation of Orion, the guide star for the Solomon Islands (latitude 7°S). Longitudinal navigation was quite another problem there being no indigenous method known for islanders to accomplish this task by celestial observations. While Old World navigators employed timekeeping mechanisms for sailing in the east-west direction, the people of Oceania used a knowledge of wind and oceanic currents, i.e., purely natural forces, to circumvent the problem successfully.

A hidden similarity has emerged from our brief glimpse at astronomical systems among indigenous tropical cultures. We contend that the horizon and zenith seem to be the underlying principle of the symbolic and organizational expressions of every system we have discussed. In the higher latitudes this is not always the case. This fundamental difference between tropical and non-tropical astronomies seems to be geographically rather than culturally determined. The debate about the degree to which environment affects the culture of man has deep and widespread roots in the 19th and 20th century history of anthropology. The old notion of the uniform evolution of all people through successive stages of sophistication from savagery through barbarism to civilization was discredited at the turn of the century by anthropologist Franz Boas and others, who showed that people most diverse in culture and language often are found living under the same geographic conditions. Field anthropologists who applied Boas's historical method in great depth to the study of diverse peoples of the world demonstrated that any explanation for similarities had better be laid to the diffusion and slow mixing of cultures rather than to independent invention.

Caught up in the success of historical inquiry, anthropologists today might tend to overlook the subtle effects of the geographical environment on cosmic views. Such effects might be discovered through application of the older comparative method of looking at specific traits among disparate cultures. But as Boas himself suggested, once culture history had begun to become established as a valid discipline, as it has now for more than half a

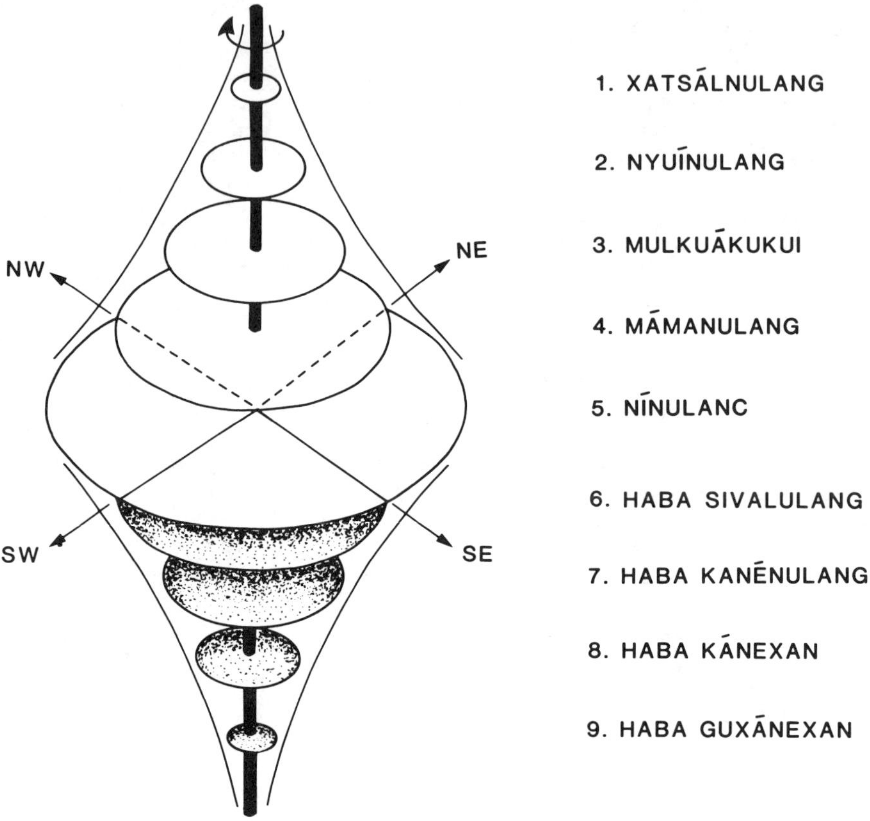

KOGI CONCEPTION OF THE COSMOS

The Kogi Indians of Colombia organize their cosmos around a vertical axis. They imagine that the universe is spindle-shaped, with nine levels. The central level is the disk of the earth, and its orientation is defined by the intercardinal directions of solstice sunrises and sunsets and link the horizon to the primary world axis. (Griffith Observatory, after Gerardo Reichel-Dolmatoff) century, the comparative method might once again reach fulfillment. It would then rest upon a broader and sounder basis in fact. The time to contrast the cosmological systems of ancient civilizations seems now to have arrived.

Traditionally, the astronomies of the ancient civilizations that I have discussed often are regarded as primitive and unscientific. The verdict in this judgement is reached because, like American anthropologists before Boas, we tend to set up our own system as the standard toward which all civilizations, technical or not, ought to proceed. Frequently, technology and precision are viewed as the solitary key to understanding and explaining natural phenomena. But the concepts of navigation in Oceania and astronomical building alignments in Mesoamerica bear little relation to the astronomical pursuits we retrieve from the Classical world and then modify and mold to shape our own modern culture. The Polynesians were driven toward practical ends while the Maya embedded their astronomy deeply within their religious attitudes and practices. Because many of these astronomies are entrenched in different cosmic perspectives, asking other questions and seeking other goals, the issue of whether the Maya astronomers were better than Greek astronomers, or indeed whether either were true scientists like us, is of little interest.

Finally, one danger in formulating any generalization lies in the ultimate expression of that generalization as a dictum. The vagaries and complexities of cultural development serve as ample warning in the present case. Of course, we might find cultures in the higher latitudes that employ the horizon in their practice of astronomy. We need look no farther than Stonehenge, where the evidence (latitude 50°-60°N) affords an excellent counterexample to my thesis, albeit based solely on evidence from alignment studies. The Maori of New Zealand, also located well out of the tropics at latitude 40°S, used a navigational system like the Polynesians', although this habit probably diffused from the tropics. Conversely, the ecliptic and celestial equator are found to surface as astronomical concepts utilized among tropical people. Indeed, pages 23-24 of the Paris Codex, an ancient Maya manuscript, have been interpreted as a zodiac consisting of 13 constellations. The Kogi tribesmen of northern Colombia and the Dogon of central Africa also pay attention to the movement of the sun among the stars. Rather than searching out the paucity of similarities between tropical and temperate astronomies, I have chosen to emphasize the differences which I find to be rather profound and widespread. The evidence as we have assembled it thus far points us toward the conclusion that once ancient people gazed at the heavens and watched sublime motion in the skies unfold, their eyes told their minds what to do.

7. Egyptian Astronomy: A Tale of Temples, Tradition, and Tombs

Along 800 miles of Egyptian Nile may be found the tatters of a civilization so taken with the sky, the imprint of astronomy remains indelible even after 5000 years. To the Egyptians the sky was a rich metaphor. Its patterns of life, growth, death, and rebirth were stenciled onto the earth. The cycles of human life resonated with the sky.

Egyptian astronomy was inextricably entwined with Egyptian religion. The Egyptians plucked images from the sky and incorporated them into their temples, tombs, and tales. New Kingdom temples, at Karnak and Abu Simbel, appear to be aligned with the sun. Paintings in the royal tombs of Thebes have numerous astronomical connotations. Even Giza's Great Pyramid is linked with the sky.

A more practical product of the Egyptian astronomers was a solar civil calendar, based upon the length of the tropical year. This scheme, and the Egyptian 24-hour system of timekeeping, evolved into the calendar and clock we use today. Curiously, these two vital components of organized society are most apparent in the paraphernalia of the priests and the dead. What we know of Egyptian astronomy is, in large measure, what we find in temples and burials.

We see in Egypt the applications of astronomy. The astronomy itself is only implied, and the monuments suggest Egyptian astronomy was a lightweight. There is no evidence of a technical vocabulary. There is no written record of systematic observations. The comparison is usually drawn to

The winged sun disk is one of the better known celestial symbols of ancient Egypt. Here, in the gap between the two towers of the pylon of the Temple of Khonsu at Karnak, it mimics the rising sun as it emerges from behind the mountainous horizon. We know that this meaning was intended by the Egyptians, for the word for the pylon, *akhet*, means "luminous mountain horizon of heaven." (E.C. Krupp)

Mesopotamia, where astronomy was mathematical and bordered on true science. This is misleading, however, for mathematical astronomy in ancient Babylonia may have begun as late as the Seventh Century B.C. By contrast, when we first think of Egypt, anything from the Old Kingdom pyramids of the Third Millennium B.C. to the Ptolemaic temples, about 2500 years more recent, may come to mind. If we are really intent on evaluating relative expertise, we must be clear about what Egypt and what Mesopotamia we have in mind.

Whether astronomy in Egypt was mathematical or scientific in a sense hardly matters. Astronomy percolated through Egyptian culture, and to appreciate its flavor we have to understand how the Egyptians interacted with the sky.

Egyptian astronomers were timekeepers. It was their function to monitor the sky to ascertain the hours and keep the calendars. We have one very late text, the Papyrus Carlsberg 9, which seems to have been written by an anonymous astronomer. It provides a table and technique for calculating the date, in the Egyptian civil calendar, of the start of each lunar month, in a 25-year cycle. The document is no earlier than 144 A.D., but Richard A. Parker, a specialist on Egyptian calendrics, has shown the text reflects a tradition 500 years older than the papyrus itself.

Hieroglyphic inscriptions are our most direct sources of Egyptian astronomy, but little else is as explicit as Papyrus Carlsberg 9. Before Jean François Champollion cracked the hieroglyphic code in the early Nineteenth Century, classical commentaries, often unreliable, defined our notion of Egyptian astronomy. Once hieroglyphics could be read, at least in part, inscriptions were copied, collected, and translated by a variety of scholars and travelers. In an ambitious and successful expedition (1842-1845) Karl Richard Lepsius, a German, acquired enough material from Egyptian monuments to attempt, in 1849, a chronology of Egyptian history. Lepsius also recognized and translated the astronomical character of many texts.

Some astronomical texts were, in fact, applied later to the problem of Egyptian chronology. By

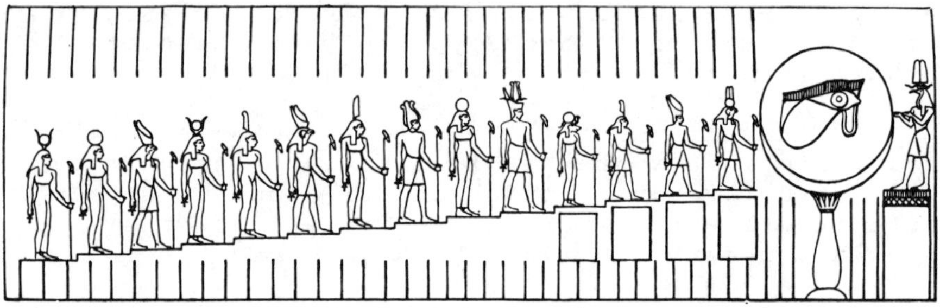

The 14 days of the waxing moon, depicted as gods, ascend ever closer to the full moon--a disk bearing the eye of Horus--enthroned in the crescent of a support stand and attended by the god Thoth (far right). An accompanying text says "the moon-eye is greeted by the sun-eye," on the fifteenth day, a poetic description of the full moon rising as the sun sets. Although this relief from the ceiling of the hypostyle hall of the Temple of Hathor at Dendera is relatively late (Ptolemaic), it preserves a truly ancient tradition: lunar festivals that were scheduled according to the observed phase of the moon. (after Budge, The Gods of the Egyptians)

associating an event mentioned in an ancient inscription with a calculated date, the ancient chronology--what happened when--was tied down. References to heliacal, or first predawn, risings of Sirius, a star of particular importance to the Egyptians, were linked by their contexts to specific dates.

Heinrich Brugsch, also a German and a renowned Egyptologist, translated many inscriptions and collected them into a five-volume Thesaurus Inscriptionus Aegyptiacarum. The first number in the series, Astronomical and Astrological Inscriptions on Ancient Egyptian Monuments, was published in Leipzig in 1883 and remained for many years the fundamental work on Egyptian astronomical texts. More recently, other researchers, among them Alexander Pogo, Otto Neugebauer, and Richard A. Parker, have analyzed these and other texts in much greater detail. Our comprehension of Egyptian astronomy has improved accordingly.

It was obvious, by Brugsch's time, that the Egyptian calendar and, in particular, the New Year were calibrated by the reappearance of Sirius in the predawn sky. Sirius, like most stars, spends part of the year--70 days, in fact--in the daytime sky. When the earth moves in its orbit about the sun, the sun appears to move eastward relative to the background stars, which are, of course, invisible. Eventually the earth will travel far enough to permit Sirius, or any other temporarily invisible star, to rise just long enough before the sun to appear, at least briefly, in the predawn sky.

The predawn reappearance, or heliacal rising, of Sirius was obvious to Egyptian astronomers. Sirius is the brightest fixed star in the sky, and its heliacal rising occurred close to the summer solstice and preceded the all-important Nile flood. The Egyptians counted three seasons and based them upon the behavior of the Nile. The seasons: "Flood", "Emergence", and "Low Water", each comprised four lunar months. Ancient texts provide us the names of these months, and reliably timed feasts occurred in each of them.

Although the Egyptian year was bound to the Nile, its months, if unattended, could quickly outpace the seasons of the tropical year. A "year" of twelve lunar months is 354 days, approximately 11 days short of the solar year. The early Egyptians solved this problem, however, by intercalating a month whenever the heliacal rising of Sirius occurred in the last third, or 11 days, of the last lunar month. We know, in fact, the name of even the intercalary month: *Thoth*, and there is evidence for use of this technique at least as early as the Age of the Pyramids.

Early in the Third Millennium another calendar was introduced. Administration of government and commerce presumably stimulated a more regularized system based on a solar tropical year of 365 days. The old lunistellar calendar continued to determine the liturgical activities, at least for awhile. Civil matters became the jurisdiction of the new solar calendar. The true length of the tropical year is closer, however, to 365¼ days. With such a discrepancy, the civil year's original synchronization with the lunistellar calendar dissolved, and a new lunar calendar was inaugurated. Its intercalations were timed by the civil new year. Even so, the civil year slowly drifted through the seasons, with a period known as the Sothic cycle, perhaps 1461 Egyptian civil years long. Rather than abandon any of the three calendars, the Egyptians maintained them all, even into Classical times.

The civil year was a sensible innovation. Its "months" were not months at all, for they were not really determined by the phase of the moon. Instead they were set to be simply 30 days each. Twelve 30-day months plus an additional five "epagomenal" days at the end generated a tropical year whose passage could be timed with the heliacal rising of Sirius. We know there were Egyptian astronomers. Someone had to watch that star.

Each of the "months" in the civil year was subdivided into three 10-day intervals called *decades*. Just as the predawn reappearance of Sirius signalled the new year and the start of the first decade of the first month, a sequence of stars, known as *decans*, was chosen to indicate the commencement of each of the other decades of the year. These decans were used also to indicate, by their risings, the various hours of the night.

Months of the old lunar calendar are represented as gods in the "Month Table" of the Ramesseum, a New Kingdom temple at Thebes built by Ramses II. In carved relief, it occupies one of the registers on a ceiling in one of the temple's halls. This detail portrays the month Phamenoth as a jackal and the following month, Pharmuthi, as a cobra-headed goddess. Another figure, an attendant, stands between them. (Robin Rector Krupp)

Julius Caesar assigned an Egyptian, Sosigenes, to convert Rome to the solar civil calendar, and through Rome the calendar became our own. We take the calendar for granted today, and yet it's a small miracle that two people can make an appointment and meet in the same place at the same time on the appointed day. Contrary to an often repeated claim, a systematic calendar is not so essential for successful planting and harvesting. This is, by now, a rather tired explanation for the early development of astronomy. The distribution of goods and services and the interaction of society are different matters altogether, however. The calendar's real power is the organization and stability it imposes on human activity. The calendar is a fundamental component of culture, and we have the Egyptians to thank for ours.

Egyptian astronomers left as a second legacy the 24-hour day. Timekeeping, like the calendar, is a basic tool of organized society. The Egyptians, we know, marked the passage of the night by the risings of stars. These were the same 36 decans whose heliacal risings announced the year's decades (plus some additional stars for the epagomenal days).

The special significance of Sirius prompted the Egyptians to choose as decans stars which mimicked the period of invisibility and pattern of reappearance of Sirius. Of course, at any given time of year only a certain number of decans would rise during the course of a night, but each, as it rose, indicated the start of another "hour". It might be thought, with 36 decans around the sky, one half of them, or 18, might be observed to rise during the night. This in turn would suggest an 18-hour night. Actually, however, the length of complete darkness varies seasonally. The Egyptians calibrated their calendar at about the time of summer solstice, when nights are shortest and only 12 decans appear.

At first, the Egyptian "hours" varied in duration from one time of year to another. Later, however, hours of equal duration were instituted, and a symmetrical set of 12 daytime hours (including morning and evening twilight) was introduced. The entire scheme is a direct consequence of the solar civil year, its decimal (10-day) subdivisions, and its calibration by the heliacal rising of Sirius in midsummer at the Nile's flood. The Egyptians made these choices, all very sound and consistent and, to them, meaningful, and we as a result entertain a 24-hour day.

Our earliest evidence of a 12-hour night dates from Egypt's Middle Kingdom. On the undersides of lids of some coffins belonging to the Ninth through Twelfth Dynasties (2160-1786 B.C.) may be seen an arrangement of decan names in a grid-like array. These designs have come to be known as diagonal star clocks after the reappearance of decans' names in one column to the next, along diagonal lines.

The diagonal star clocks are not clocks at all but simply diagrams, or tables, which indicate what star marks each hour of the night during each of the 10-day intervals. The earth's gradual movement around the sun shifts each decan's position, and therefore time of rising, with respect to the sun. What the *imy-wnwt*, or "hour-watcher", saw at the twelfth and last hour of the night will rise at the eleventh hour ten days later. Each decade the star slips back an "hour", and its name leaves a diagonal trace on the coffin lid.

By the era of the Ramesside pharaohs (1300-1100 B.C.) the watching of decans had shifted from risings to transits. Our evidence is still found in a funereal context, in this case, a diagram of star names and positions shown in association with a seated figure. These star clocks appear in the royal tombs of the Ramessides.

Missing, of course, are the actual papyri, if indeed they existed at all, used by the Egyptian astronomers to keep track of which stars did what when. We should not underestimate, however, the importance of the decanal system. Inclusion of these star clocks into the environment of the dead suggests knowing the hour and the date was as vital for the transit to the next life as for the passage through this. Prayers or charms--a right word at the appropriate time--protected the dead.

The sky's drama was equated with human destiny. We can get the flavor of this in the inter-

A carefully selected set of stars was used to monitor the passage of the hours and the days of the year. These stars were known as decans, and Sirius was the most important of them. Sometimes the decans show up in inscriptions in New Kingdom tombs. Identified on grids known as Ramesside star clocks, the decans in the tombs were intended to ensure that a proper prayer or charm would be invoked at the proper time. Like the star clocks, the prayers also were painted onto the walls and ceilings of the chambers for the dead kings. (tomb of Ramses VI, E.C. Krupp)

changeability of astronomical language and funereal language. When a decan star disappears for 70 days (as it was selected to do), it is said to die and enter the embalming house of the underworld, Duat. It remains there for 70 days and loses its impurities. Then, it is reborn in the east. Similarly, a body was embalmed for 70 days prior to burial. The underworld itself was divided into 12 regions, one for each hour of the night. It was the sun's destiny to enter Duat at sunset, pass through the perilous realm, and emerge victorious at dawn. Accounts of this symbolic journey appear as religious texts in the royal tombs of Thebes. We see there as well portrayals of the sky goddess, Nut. The sun disk at her mouth symbolizes sunset. She swallows the disk, and it passes through her body. For the dawn, it reappears at her loins, literally born from the sky. The daytime disappearance and nightly rebirth of the stars is shown in the same way.

One of the most persistent astronomical images of ancient Egypt is the Northern Group. This unlikely collection of celestial residents includes, among others, a bull (or often just its foreleg), a hawkheaded human, sometimes shown with a spear, and an upright female hippopotamus, sometimes cloaked with a crocodile. In some portrayals the hippopotamus leans upon an enigmatic device. It looks like a knife, is called a mooring post, and represents the north pole of the sky. The stars of the Northern Group are the circumpolar stars. To the Egyptians they were the "imperishable" or "undying" stars, for they never passed below the horizon into Duat.

What we today call the Big Dipper, or Plough, was the bull (or its foreleg). Its circuit about the pole was symbolized by the tethers which continued down to the mooring post. The hippopotamus was composed, at least in part, of some of the stars of Draco the Dragon, and Thuban, the star nearest the pole in the Old Kingdom, may have been associated with the "mooring post".

Although much has been made of the Great Pyramid's supposed symbolic significance, the "imperishable stars" help confirm it was, indeed, the tomb of Khufu, the Fourth Dynasty pharaoh who built it.

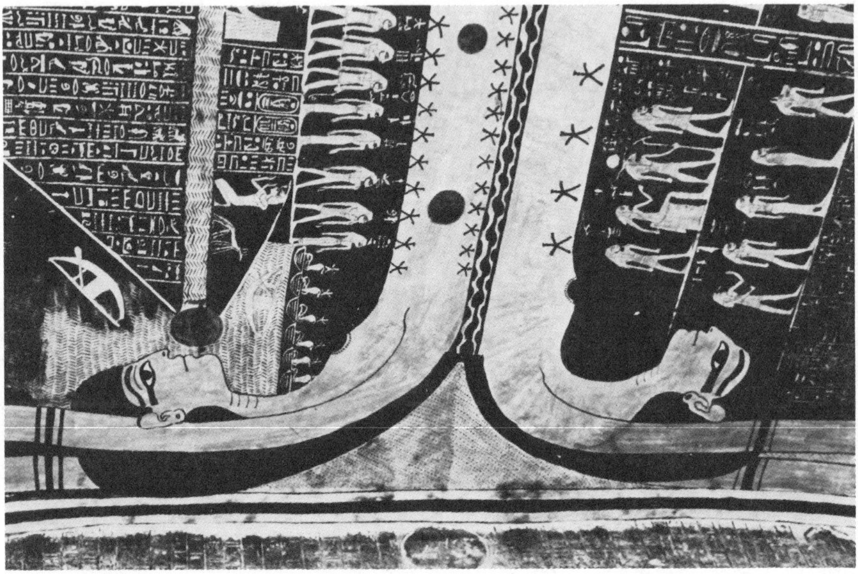

Nut, the Egyptian goddess of the sky, appears twice on the ceiling of the tomb of Ramses VI in the Valley of the Kings. A sun disk is poised before the lips of the Nut on the left. She swallows it, and it passes through her star-studded body. The left side therefore represents the sunset and the night, when the sun is consumed by the horizon, and the inscriptions on this side are texts from the Book of the Hours of the Night. Stars alone pass through her body on the right. This refers to the day, when the stars are "swallowed," and the texts come from the Book of the Hours of the Day. (E.C. Krupp)

Stars near the north celestial pole comprise the Egyptian "Northern Group" of constellations, depicted here on the ceiling of the burial chamber of the tomb of Seti I, in the Valley of the Kings. The hippopotamus is leaning upon a "mooring post," which is the pole, or pivot, of heaven. Tethered to it is a bull. For us, the bull's stars are those of the Big Dipper, which circles endlessly around the pole, tethered just as the bull is. (E.C. Krupp)

The Great Pyramid of Giza is accurately aligned to the cardinal directions. Inside, an odd assortment of passages and chambers has prompted extravagant speculation about the monument's meaning and use. (Robin Rector Krupp)

At the heart of the Great Pyramid is only a stark room, empty save for the open sarcophagus on its west side. The room's walls and ceilings are built of massive blocks of granite, quarried and carried from Aswan, 500 miles to the south. Here we look toward the room's northeast corner, where the low entrance to the room from the Grand Gallery is visible. On the wall to the left, a small opening marks the end of the northern "air shaft," which bends upward at an angle of 31° and penetrates through the pyramid to its north face. (E. C. Krupp)

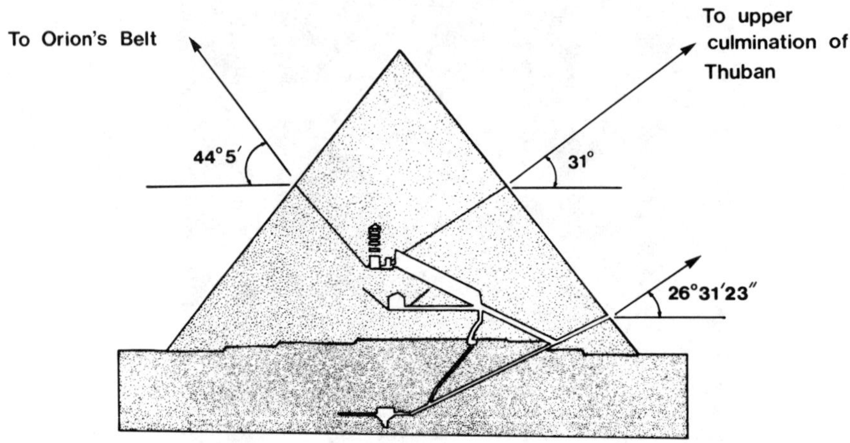

The Great Pyramid of Giza

Two shafts extend from the King's Chamber of the Great Pyramid at two different, astronomically significant directions. One points to the upper culmination of Thuban, which was close to the north celestial pole at the time the Great Pyramid was constructed. The other heads for the belt of Orion. The north target symbolizes the "undying" circumpolar stars. The south target is a celestial form of the god Osiris, who presided over resurrection and renewal. Both celestial destinations are appropriate for the pharaoh as he passes from this life to the next. (Griffith Observatory)

We know that the sides of the Great Pyramid are aligned with great accuracy to the cardinal directions. These, after all, are astronomical, for they derive their meaning through the apparent daily rotation of the sky. The worst error is only five and a half arcminutes, just about one sixth the size of the full moon as we see it in the sky. All the more remarkable is the maintenance of this accuracy in an artificial mountain of 2,300,000 stone blocks ,756 feet long at the base on a side and originally 481 feet high.

Inside the Great Pyramid is a mystifying collection of chambers and passages. One of these, the "King's Chamber," still contains the empty, lidless remains of a granite sarcophagus. Two curious openings may be found just a few feet above the floor and toward the east end of the room, one on the north wall and one on the south. They are the ends of two shafts. Both extend horizontally and then bend upward. The north shaft is set at an angle of 31 degrees, the south at 44 degrees 5 minutes. The shafts continue at these angles nearly to the outer faces of their respective sides before they bend horizontally again and open to the sky.

Ventilation has been the usual explanation for the King's Chamber "air shafts", but the Egyptians were better at air conditioning than this. And in any case, this is a very difficult way to achieve the presumed effect. Egyptologist Alexander Badawy of U.C.L.A. assisted by an astronomer, Dr. Virginia Trimble of the University of California, Irvine, demonstrated in 1964 that the north shaft pointed toward the upper culmination of Thuban's circumpolar arc in 2650 B.C., when the Great Pyramid was built. Similarly, the south shaft was oriented on the transit of Orion's belt, perhaps on its central star, Alnilam. These astronomical alignments need not be ascribed to chance, for Badawy cited written evidence to explain them.

Inside the pyramid of Unas (Fifth Dynasty, 2350 B.C.) at Saqqara, the walls are completely carved in hieroglyphics. These writings are the Pyramid Texts, and they describe, among other things, the pharaoh's destiny. He is to ascend to the imperishable stars and manage their appointed rounds. He is to ascend to the sky to join Osiris, the god of resurrection, as well.

The Great Pyramid's northern shaft was not a sighting tube for Thuban. It symbolized the ascent of the pharaoh to the realm of the undying stars. In the same way, the southern shaft represented the pharaoh's journey to Osiris. Osiris, in the Egyptian sky, was the constellation Orion. Astronomy, funeral rites, and religion were combined in the Great Pyramid on monumental scale.

Astronomical allegory is a fundamental constituent of Egyptian religion. Some temples are covered with astronomical inscriptions and imagery. At Dendera's Temple of Hathor the New Year Festival was celebrated on the roof in an open kiosk. The temple is Ptolemaic and rather recent (125 - 30 B.C.), but some of the traditions its walls preserve are much older. At Dendera the heliacal rising of Sirius was dramatized by allowing the light of the rising sun to fall upon a statue of Isis, the goddess associated with Sirius. The long stairway on the temple's east side leads to the roof, and its walls portray the New Year's procession. Among the numerous astronomical inscriptions at Dendera is this symbolic description of the heliacal rising of Sirius (Isis) in the company of the sun (her father).

> "Radiantly, above her father's
> forehead, the golden-one rises,
> and her mysterious form occupies
> the bow of his solar boat."

Other texts at Dendera explicitly describe the behavior of Sirius and the Egyptians' perception of her role.

> "Her rays unite with the rays of
> the luminous god on that beautiful
> day of the sun disk's birth on the
> morning of the New Year's feast."

Sirius is

> "a feminine sun which appears in
> heaven at the beginning of the year
> as a divine star, Sothis, the queen
> of the decan stars, whose rays light
> the earth like the morning sun. She
> is the mistress of the year's begin-

Sirius appears here on the left as the goddess Isis. She sails in her celestial boat upon the ceiling (or "sky") of the burial chamber of the tomb of Seti I, and she is accompanied by Orion. He is portrayed as Osiris, and he leads Isis across the heavens just as Orion precedes Sirius over the course of the night. (E.C. Krupp)

ning, and entices the Nile out of its
source hole to provide life to the
living people."

Abundant evidence documents the astronomical connotations of Egyptian temples, but the sky's theatrics may have even been staged in a temple's layout and orientation. Sir Norman Lockyer, the renowned astrophysicist and founder-editor of Nature, was the first to argue that Egyptian temples had been precisely aligned on the risings and settings of the sun and bright stars. His measurements and conclusions were published in 1894 in The Dawn of Astronomy. The quality of the work was mixed. Egyptologists ignored it or regarded it unfavorably.

In 1973, Dr. Gerald S. Hawkins, a British-born astronomer known, like Lockyer, for his astronomical interpretation of Stonehenge, approached the problem of New Kingdom temple alignment more modestly. His results are more convincing.

Karnak, near modern Luxor and 450 miles south of Cairo, is the site of the Great Temple of Amun-Re. The temple is huge and is itself but a part of a gigantic complex of temples. Construction and modification of the Great Temple of Amun-Re continued over several centuries, but most of the important building belongs to the Eighteenth Dynasty (1558-1303 B.C.). With a northwest-southeast orientation, the temple was thought targeted, by Lockyer, on the summer solstice sunset. Hawkins by contrast has shown that a winter solstice sunrise alignment is more likely. He associates this astronomical event with a solar sanctuary possessing a window that opened to the midwinter sunrise in the southeast. The sanctuary's axis is aligned with that event, and Hawkins cited some metaphorical texts to support the idea. The principle of Karnak's "High Room of the Sun" is duplicated on a smaller scale 300 miles farther south in the rockhewn temple of Ramses II at Abu Simbel. There the main temple appears to have been directed to the sunrise on the civil new year on the occasion of the thirtieth year jubilee of Ramses's reign. This date coincides with 18 October in our calendar. A small, easily ignored little chapel is situated at the north end of the main temple's monumental facade. Architecturally this chapel resembles the

The Temple of Hathor at Dendera was the site of the New Year's Festival, and this appears to have been its primary function. Here, on the north side, the entrance provides access to the spectacular hypostyle hall. Its ceiling is completely devoted to astronomical symbols. (Robin Rector Krupp)

The New Year's procession at Dendera would climb to the roof of the Temple of Hathor by a long stairway. Reliefs on the walls of the stairway still illustrate the procession. (Robin Rector Krupp)

On the roof of the Temple of Hathor at Dendera, participants in the New Year ritual observed the heliacal rising of Sirius from this open kiosk. (Robin Rector Krupp)

Sir Norman Lockyer thought that the main axis of the Great Temple of Amun-Re at Karnak was intended to conduct the light of the summer solstice sunset far into the sanctuary along this line-of-sight. The high wall is the temple's outermost pylon. (E.C. Krupp)

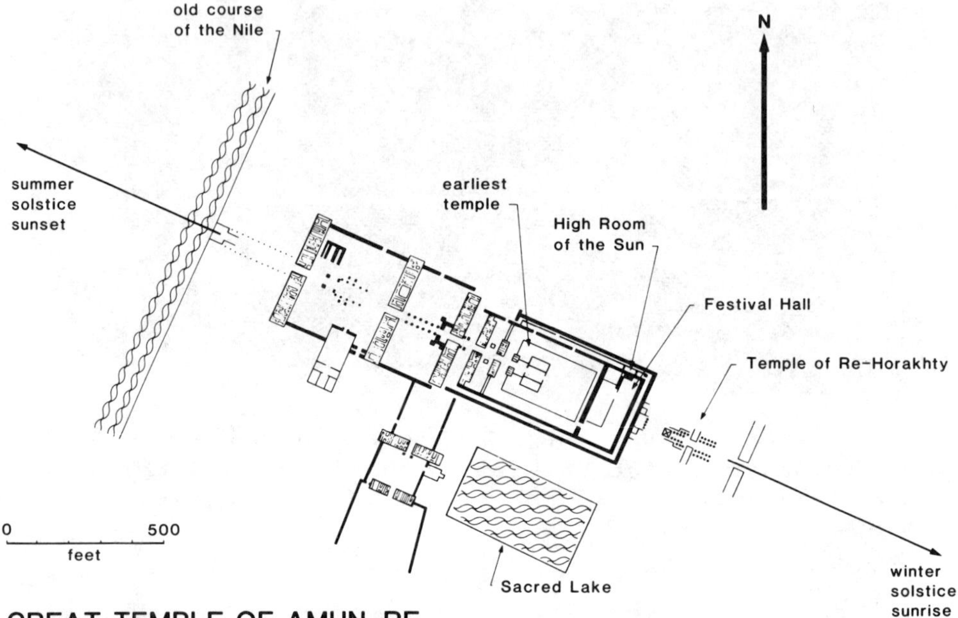

GREAT TEMPLE OF AMUN-RE
KARNAK

Several possible alignments at Karnak were proposed by Lockyer, and two involved the sun. To the northwest, the Great Temple of Amun-Re faced the summer solstice sunset, and, in the opposite direction, was the winter solstice sunrise. Temples opening to the southwest were, in fact, built beyond the back wall of the main temple. (Griffith Observatory)

This, according to Lockyer, was the view to the summer solstice sunset from the sanctuary of the Great Temple of Amun-Re and along its axis to the northwest. (E.C. Krupp)

Astronomer Gerald S. Hawkins pointed out the possible astronomical significance of this modest room toward the northeast corner of the Great Temple of Amun-Re. The bottom of the ruined stone frame of a window appears just beyond a massive stone block in the foreground. The block is an altar, in the shape of four *hetep* glyphs. *Hetep* means "offering," and the room seems to have served a ritual purpose. Hawkins calls it the "High Room of the Sun." (Robin Rector Krupp)

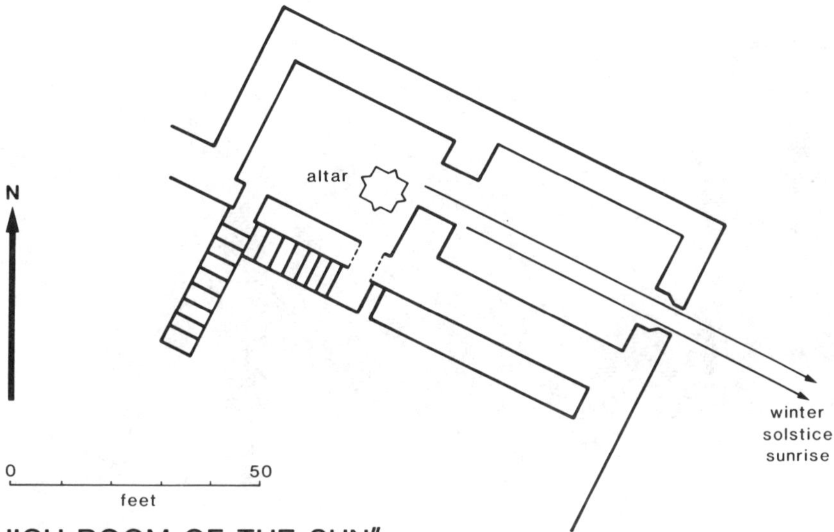

"HIGH ROOM OF THE SUN"
CHAPEL OF RE-HORAKHTY
TEMPLE OF AMUN-RE
KARNAK

Hawkins's analysis suggested that the High Room of the Sun was open to the winter solstice sunrise and oriented toward the southeast. It was not necessarily an observatory but, more likely, a sanctuary for ceremonial observation of the sun at a key time in its annual cycle. (Griffith Observatory)

Ramses II was responsible for the amazing rock hewn temple of Abu Simbel. Its main entrance is beyond the ramp, between two of the gigantic statues of Ramses seated. To the right, the facade of a small, relatively inconspicuous chapel mimics the front pylon of more monumental temples. (Robin Rector Krupp)

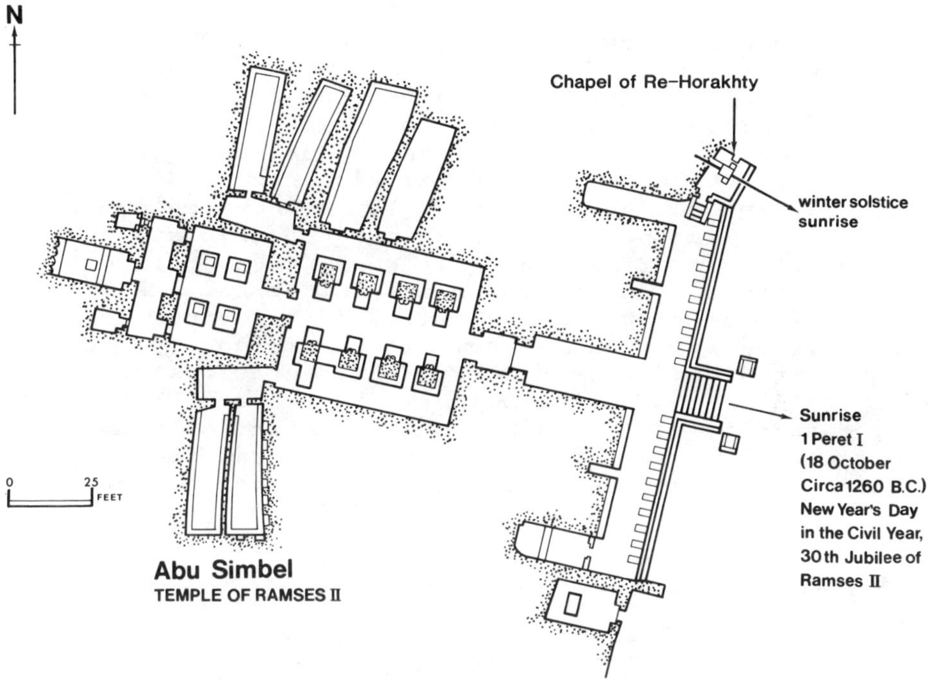

The main axis of the Temple of Ramses II at Abu Simbel coincides with the sunrise direction for New Year's Day, in the "wandering" civil calendar of ancient Egypt, during the time of Ramses II's Jubilee. This was an important time for the kingdom and the pharaoh, and it was associated strongly with the idea of renewal. Before the temple was lifted to escape the rising waters of Lake Nasser, the sun first appeared in a distant horizon notch. This, in fact, is the inspiration for the temple pylon profile and for the akhet glyph itself, which looks like a disk cradled in the rounded arms of a thickly padded couch. (Griffith Observatory)

High Room of the Sun, and it is skewed to the winter solstice sunrise.

We need not doubt the existence of ancient Egyptian astronomers. They assume, in fact, increased importance as we realize how deeply astronomy's applications penetrated Egyptian life. We relive their achievements in our calendars and clocks. There is little direct evidence of the astronomers themselves, however. Amenemhet, of the Sixteenth Century B.C., is the earliest, and one of the few, we know by name. It is the tomb of Nakht, an astronomer of the Eighteenth Dynasty, however, that offers the most convincing proof of the presence of an astronomer's lifestyle. The royal Theban tombs in the Valley of the Kings are rich in astronomical associations and imagery: the Northern Group, star clocks, representations of Sirius, Orion, and the planets, and more. Nakht, by contrast, decorated his tomb with those things closest to an astronomer's heart--enthusiastic winemaking, sumptuous banquets, and women as dancers and musicians, delicate and scantily clothed.

References

Badawy, Alexander
1964 "The Stellar Destiny of Pharaoh and the So-Called Air-Shafts of Cheops' Pyramid," Mitteilungen des Instituts fur Orientforschung, Band X. pp. 189-206.

Brugsch, Heinrich
1883 Astronomical and Astrological Inscriptions on Ancient Egyptian Monuments. Leipzig: J.C. Hinrich. (English translation serialized in *The Griffith Observer*, published by Griffith Observatory in Los Angeles, 1978-1979).

Hawkins, Gerald S.
1973 Beyond Stonehenge. New York: Harper & Row.

1974 "Astronomical Alignments in Britain, Egypt, and Peru," Philosophical Transactions of the Royal Society of London, vol.276, no.1257 ("The Place of Astronomy in the Ancient World," ed. F.R. Hodson), pp. 157-67.

1975 "Astroarchaeology: The Unwritten Evidence," in A.F. Aveni, ed., Archaeoastronomy in Pre-Columbian America. Austin: University of Texas Press, pp. 131-62.

1975 "Stargazers of the Ancient World," 1976 Yearbook of Science and the Future. Chicago: Encyclopaedia Britannica, pp. 124-37.

Krupp, E.C.
1978 (March) "Great Pyramid Astronomy" The Griffith Observer (Griffith Observatory, Los Angeles) vol. 42, no. 3, pp. 1-18.

Krupp, E.C. (ed.)
1978 In Search of Ancient Astronomies. Garden City, New York: Doubleday & Company, Inc.

Krupp, E.C.
1979 "Ancient Watchers of the Sky," 1980 Science Year. World Book Science Annual. Chicago: World Book--Childcraft International, Inc. pp. 98-113.

Krupp, E.C.
1981 "Recasting the Past: Powerful Pyramids, Lost Continents, and Ancient Astronauts," Science and the Paranormal (ed. George O. Abell and Barry Singer). New York: Charles Scribner's Sons, pp. 253-295.

Krupp, E.C.
1981 "The Sun Gods," Fire of Life: The Smithsonian Book of the Sun. (ed. Joe Goodwin et al). Washington, D.C.: Smithsonian Exposition Books, pp. 160-167.

Krupp, E.C.
1983 Echoes of the Ancient Skies. New York: Harper & Row, Publishers, Inc.

Lockyer, J. Norman
1973 The Dawn of Astronomy (1894). Reprint edition, Cambridge, Massachusetts: M.I.T. Press.

Neugebauer, O.
1962 The Exact Sciences in Antiquity. New York: Dover Publications, Inc.

Neugebauer, O.
1975 A History of Ancient Mathematical Astronomy Part 2. New York: Springer-Verlag.

Neugebauer, O., and R.A. Parker
1960 Egyptian Astronomical Texts I. The Early Decans. Providence, Rhode Island: Brown University Press.

Neugebauer, O., and R.A. Parker
1964 Egyptian Astronomical Texts II. The Ramesside Star Clocks. Providence, Rhode Island: Brown University Press.

Neugebauer, O., and R.A. Parker
1969 Egyptian Astronomical Texts III. Decans, Planets, Constellations and Zodiacs. Providence, Rhode Island: Brown University.

Parker, R.A.
1974 "Ancient Egyptian Astronomy" Philosophical Transactions of the Royal Society of London Vol. 276, no. 1257. ("The Place of Astronomy in the Ancient World," ed. F.R. Hodson) pp. 51-65.

Parker, R.A.
1978 "Egyptian Astronomy, Astrology, and Calendrical Reckoning," Dictionary of Scientific Biography. Vol XV, suppl. I, New York: Charles Scribner's Sons. pp. 706-727.

Trimble, Virginia
1964 "Astronomical Investigation Concerning the So-Called Air Shafts of Cheops' Pyramid. Mitteilungen des Instituts fur Orientforschung, Band X pp. 183-187.

Index